# HOW MAN MOVES

# HOW MAN MOVES

## Kinesiological studies and methods

*By*
SVEN CARLSÖÖ
*Professor, Royal Caroline Institute, Stockholm*

Translated by
William P. Michael, B.A. (Hons.), F.S.C.T.A., M.A.T.A.

HEINEMANN : LONDON

William Heinemann Ltd
15 Queen St, Mayfair, London W1X 8BE

LONDON   MELBOURNE   TORONTO
JOHANNESBURG   AUCKLAND

434 90215 2

Original version: *Människans rörelser*, published in Stockholm by PA-rådet, the Swedish Council for Personnel Administration; revised and enlarged for this translation.

Photoset and printed in Great Britain by
BAS.Printers Limited, Wallop, Hampshire

# Preface

The movements of man have always given rise to speculation and discussion and were even studied in ancient times. But only in our modern day and age, the last few decades in particular, have there been any systematic attempts to study the laws governing human motion using the methods of modern science with their demand for cogency and objectivity.

The word kinesiology has been formed from the Greek words *kinesis*, meaning movement, and *logos*, a combining form used in the names of sciences. It means the science of human and animal movement.

Kinesiology is an inter-disciplinary science, primarily based on anatomy, physiology, and mechanics but also dealing with orthopaedics, neurology, psychology, ergonomics, sports medicine, and the theory of gymnastics. Our presentation will emphasize the anatomical, physiological, and bio-mechanical aspects.

Naturally, we shall only take up the most important elements in human movement. The presentation will largely consist of lectures in kinesiology given by the author in recent years for students of medicine at the Royal Caroline Institute and at the College of Physical Training, both in Stockholm, Sweden. Physical therapists, vocational therapists, ergonomists, and odontologists have also become acquainted with kinesiology as presented here through refresher courses.

As a rule, students were sufficiently familiar with anatomy, physiology, and mechanics to follow the lectures without difficulty. And it is assumed that the reader will also be in possession of up-to-date knowledge in these disciplines or will be able to freshen up their knowledge with supplementary study.

## Preface

A glossary of terms and an index may be found at the rear of the book and references have been provided for the convenience of readers wishing to do further study of kinesiology.

*Stockholm* SVEN CARLSÖÖ

# Contents

*Contents*

# 1 Development and Organization of Movement

Needless to say, these words on the development and organization of movement make no pretence of being comprehensive. Only a few phenomena will be discussed in order to show, *inter alia*, the intimate connection between the nervous system and the locomotor system and how environment and inherited characteristics collaborate in the design of movements.

The gradual development of movement begins at the embryo stage. Both anatomical and physiological development follow a scrupulously laid down timetable. Even if each embryo has its own inherited characteristics, there are only insignificant chronological differences in development pattern among embryos. Thus, on the twenty-sixth embryonic day, two small bulges appear on the sides of the embryo's body. These are the beginning of arms. Two days later, the prospective lower extremities show up as a pair of bumps.

In the second month of pregnancy, the initially shapeless, rudimentary extremities rapidly begin developing into segments. By the end of the month, these segments have acquired their characteristic shape. Thighs, knees, lower legs, ankles, hands, and fingers look like miniature versions of the corresponding organs in adults. Dimensions are obviously small here, as the embryo is only 4 cm long at the end of the second month of pregnancy. The skeleton still consists of cartilage. Calcification of the cartilagenous skeleton begins during the third month of pregnancy.

The ability to move demands the existence of both joints and muscles as well as nerves. In the sixth and seventh weeks, the first connections are established between nerves and muscles and a very active and critical period for the embryo

1

and its locomotor system then begins. This is the foetal motor sequence which now continues up to the fourteenth week of pregnancy. No true movements by the human embryo occur until after the seventh week of pregnancy.

## FOETAL MOVEMENTS

The first movements are trunk movements. The upper part of the trunk is bent forward and back. Everything suggests that these trunk movements are reflex motions triggered by pressure, sound, or contact stimuli from the foetal environment. It has been established that if the lips of a foetus are touched, the foetus responds by bending its upper body to the side and rapidly drawing back its arms. Such movements are generally called mass movements, as large parts of the body mass are set into motion, and not merely the body part stimulated.

Isolated movements of the extremities appear somewhat later. In the eighth week, the palms become sensitive and the hand closes when touched. In the ninth and tenth weeks, the number of nerve and muscle connections increases sharply and the foetal pattern of movement develops at an increasingly rapid pace. Around the fourteenth week, one may assume that all of the body's basic movements have appeared. The foetus is then able to kick, move its feet and toes, bend its knees, elbows, and wrists, clench its fists, turn its head, gape, wrinkle its brow, etc.

In the fifth month, movements become so powerful that the mother begins to feel them. There are not only simple, isolated, reflex movements but complicated movements as well. The fixed patterns of movement characteristic of man may now be discerned. By this we mean certain congenital, reflex movements which are somewhat more complicated than simple, isolated reflex movements. The fixed movements provide an assortment of functional movements which could, indeed, take place separately but which are mainly important components in most of man's movement patterns.

In the eighth month, the foetus ends up in a head-down position, probably because the head is the heaviest part of the body. Even if no major body movements take place during the last foetal month because of lack of space, the arms and legs are still quite active.

## INNERVATION AND PATTERNS OF INNERVATION

Before we begin to discuss the post-natal developments of movement in man, it might be useful to mention briefly a few experiments with animals.

In the embryos of spur-dog sharks, rhythmic, repetitive muscle contractions occur even before muscle fibres have been supplied with nerves. These spontaneous contractions are repeated without interruption until motor end-plates have been developed and nerve cells have taken over the task of stimulating muscle fibres. The first neural connection between the central nervous system and muscle is made by efferent fibres, the alpha-neurones. Afferent nerve fibres from the muscle spindles develop later. However, their development is so rapid that it is completed before the terminal organs of the efferent fibres, the motor end-plates, have been completely formed.

Efferent fibres to motor spindles, gamma-neurones, are the last part of the muscle nerve supply to be developed. In the cat, these neurones to the distal hind leg muscles begin to function only in the third post-natal week. Thus, new-born kittens lack tonic reflex movements. Since these reflexes provide the neuromuscular foundation for the standing body position, the kitten is unable to walk during the first 3 weeks of its life. Later development of this part of the proprioceptive system is partly compensated for, with regard to the kitten's orientation to its surroundings, by exteroceptive touch impulses from the skin.

Even after the nervous system has taken over control of muscle contractions, these still remain stereotyped for a time and are repeated at regular intervals. There is no systematic interplay or co-ordination before interneurones, i.e. nerve cells in the central nervous system which connect various nerves to one another, have developed. The first co-ordination to appear, at least in certain animals, is the interplay between corresponding muscles on the right and left sides. If the number of contractions and movements increase on one side, they decrease on the other.

3

In the next phase of development, this initial co-ordination mechanism is connected to nerve cells higher up which, in turn, are connected to parts of the brain located even higher up, in still more complicated systems. These systems are organized in fixed patterns for every genus and species and they stand ready to co-ordinate stimuli and musculature in a later stage of development in patterns of movement characteristic of the genus and species. This development stage appears to be of genetic rather than functional character. Movements develop quite simply as the result of the maturation process through which the nervous system and locomotor system pass. Certain innervation patterns fundamental to and characteristic of a genus, e.g. cats, dogs, Homo sapiens, etc., are built up in the central nervous system very early on.

One could regard a pattern of innervation as a complicated system of nerve cells linked together in chains passing through various parts and levels of the brain and spinal cord. Every pattern of movement has a corresponding pattern of innervation in the brain and spinal cord. The pattern of innervation is activated by, e.g., a stimulus, a touch, visual impression, or sound, and a stream of neural impulses is sent to the muscles with which movement is to be performed.

THE DEVELOPMENT OF MOVEMENT FROM BIRTH TO MATURITY

The gradual development of man's movements, beginning in the embryonic period, from primitive, isolated reflexes in the second month of pregnancy to the rather highly developed co-ordinated forms of movement in the eighth and ninth month, continues after birth. The flexor musculature is the most powerful of the still slack muscles. Therefore, a baby frequently lies in a hunched-up position with the knees drawn in towards the stomach and the arms at face level. A child, too, often lies with one arm outstretched and the head turned in the same direction; this reflexive position is called the tonic neck reflex. This reflex, triggered by proprioceptors in the neck and cervical spine, in which parts of the brain such as the cerebellum and vestibularis are involved, plays a very important role in the movements of the trunk and extremities in adults.

4

If the palm of a new-born baby is stimulated with a spoon or finger, for example, the infant's hand grasps the object strongly. This is the same grip reflex which occurs during the intra-uterine life when the four ulnar fingers grasp the thumb or when the hands grasp the umbilical cord.

No apparent functional interplay between the extremities or different parts of the body can be traced in infant movements. If face skin is stimulated by touch, the eye by a powerful ray of light, or the ear by a sound of sufficient strength, disorderly mass movements develop. Such a stimulus may simultaneously produce mimic movements, head turning, arm and leg movements, and even affect breathing movements. These reflex-induced chaotic patterns of movement will subsequently come to be controlled by higher centres of the brain, after having been subject to the spinal cord and lower (caudal) part of the brain stem. This occurs in step with the development of the sensory organs. The afferent flow of impulses from eyes, ears, touch receptors in the skin, proprioceptors in muscles, joint capsules, ligaments, and fasciae play a fundamental role in the development of locomotor organization.

A pattern of innervation, or perhaps more accurately, the basis for a pattern of innervation, for the gait is already present at birth. If you hold a new-born infant with its head upwards and feet downwards, allowing the feet to come in contact with a table top for example, this touch on the soles of the feet triggers off leg movements similar to those made by man in walking. A fully developed innervation pattern is lacking, however, until the day the child can walk. At birth, there is, *inter alia*, an inadequate proprioceptive flow of impulses. Only when the organs from which proprioception emanates, e.g. joints, muscles, ligaments, etc., have achieved a certain level of development can the innervation pattern be completed. It also seems rather likely that certain central neurones and neurone connections have to develop before a fixed, established innervation pattern can exist and the child is able to walk.

In the first years of life, locomotor development is intense and occurs at a rapid pace. Development takes place in a cranio–caudal direction and from proximal to distal joints. This conformity in the development of these patterns of movement is striking. Irrespective of race or social conditions, the sequence of movement development is the same in almost all children, even though a number of individual differences may occur.

When a 1-year old grasps an object, it uses the entire hand for the grip. It 'draws', for example, using large, sweeping arm movements. The movement arises in the shoulder while the elbow and wrist remain fixed. Only 6 months later is the child able to exploit the possibilities of the elbow. The thumb–forefinger grip is not used until the child is about 3.

Once the child around the age of 1 has learnt to stand and take its first few steps, a period starts in which development is very rapid. By moving in space with increasing freedom, the child's distance perception and his eye–hand co-ordination are improved. In every important respect, the 3-year old is the master of his body, coarse mobility in particular. It is remarkable to note how the child's muscular activity is transferred from gross movements to fine manipulative movements. New hand grips and positions are continuously induced by maturation of the neuromotor system. The extraordinary liveliness and mobility which characterizes the pre-school child contributes towards perfecting control over movements and positions.

Manual dexterity is improved up to the age of 6, after which time there is stagnation or even decline in skill. The explanation is probably to be found in changes in body proportions. The child is then in the so-called 'stretching' period during which there is relatively rapid growth in height and slow increase in weight, normally making the child thinner. The nervous system is unable to keep up with this rather rapid change in the proportions of body parts. It takes a while before the innervation pattern can adapt efferent outflow to continuously changing afferent impulses whose variation is caused by rapidly growing skeletal parts, muscles, joints, and ligaments.

Around the age of 8, this discrepancy in development has disappeared and a gentler phase of development begins which lasts until pre-puberty. From the eighth year to puberty, there is minor growth in height but powerful growth in breadth. This is usually called the second 'expansive' period. The first 'expansive' period occurs from the second to the fourth years. In the first stage of puberty, there is a sharp increase in height, the second 'stretching' period. Growth in height declines during the latter half of puberty while growth in breadth increases strongly instead, the third 'expansive' period.

According to several American studies from the 1940s, a boy's muscular strength increases successively into his twenties and a girl's up to the age of 14. Between the ninth and fourteenth year of life, muscular strength increases in boys by about the same amount each year. In the following years, i.e. up to about 17, the annual increase in muscular strength is somewhat greater each year. Thereafter, muscular strength increases relatively little. It is greatest around the age of 25. Muscular strength appears to be the physical characteristic which develops most strikingly after the age of 6. Eighty per cent of a 20-year old boy's muscular strength is acquired after the age of 6 while height only increases by 30–35 per cent.

By co-ordinated movement we mean a movement well defined in form, strength, and speed. There is no objective method for measuring a person's general co-ordination. However, a number of more or less subjective methods for appraisal have been devised. In the 1920s an American method was developed for evaluating co-ordination. Subjects had to execute certain movements and adopt certain positions requiring especially good balance. According to this test, the co-ordination of girls is greatest at 13–14 and of boys at 18–20. It is difficult to say why the development of co-ordination in girls stops so early. It seems that there is a connection between muscular strength and muscular co-ordination. But there may be other factors in the picture, e.g. different interests or altered hormone production. In the lower age groups the difference between boys and girls is insignificant. Certain co-ordination movements are made better by girls while others are done better by boys. In movements requiring agility, boys would appear to be superior.

In contrast to growth in strength, co-ordination and body balance in boys improves very slowly during puberty. Thus, co-ordination and muscle strength do not display parallel development.

7

## How Man Moves

In a German study from the middle of the 1950s, it was found that almost all 11-year old boys could run 50 metres equally fast, even though height varied from 120 to 160 cm. This was also the case for a group of 12-year olds, who could run somewhat faster than 11-year olds. Fourteen-year old boys ran faster still but height had a definite influence on competition results; the taller the boy, the better the results. Presumably, it is not body length in itself which influences speed in 14-year olds but the varying degree of sexual maturity. Male hormones are thought to exert a favourable influence on muscular strength, and it is possible that tall 14-year old boys are more sexually mature than shorter boys. Among boys having passed puberty, e.g. 18-year olds, no connection was found between height and speed. Height had no demonstrable effect on running ability in 12-, 14-, and 18-year old girls. But 14-year olds were much faster than 12-year olds, while differences between 14-year olds and 18-year olds were insignificant.

Body weight increases at the same time as height during the years of growth. Japanese, Polish, and Danish studies have all shown that while height increases from, e.g. 110 cm to 170 cm, body weight increases from 18 kg to 62 kg. Thus, height increases 1·5 times and body weight 3·4 times, i.e. increase in body weight is proportional to the cube of height increase.

The connection which may hypothetically be expected to exist between body dimensions and physical performance has been the subject of wide discussion. If one assumes that all people are geometrically similar to one another, even if size varies, all linear dimensions and all functions which are proportional to linear dimensions should be proportional to the height of each individual, all areas and all functions proportional to areas should be proportional to $(height)^2$, and all volumes and weights and functions proportional to these values should be proportional to $(height)^3$.

Such an assumption would mean that all length measurements such as arm and leg length and the moment arms of the muscles of two boys of the same age who are respectively 170 cm and 190 cm tall would be as 170:190, i.e. 1:1·12, all area measurements (e.g. muscle cross-sectional area) as $170^2:190^2$, i.e. 1:1·25, and all volume and weight measurements (e.g. muscular mass and body weight) as $170^3:190^3$, i.e. 1:1·39. Since muscular strength, when all else is equal, is proportional to the muscle's cross-sectional area, the tall boy's muscles should be capable of producing 1·25 times more tension than the short boy's muscles. This would be true, for example, in lifting different external objects but not in lifting one's own body. Muscular strength is indeed 1·25 times greater in the tall boy but his weight is also 1·39 times greater. The tall boy is, thus, handicapped in climbing. In running and jumping, speed and the height to which the body's centre of gravity can be lifted are proportional to muscular force multiplied by the muscle's shortening distance but inversely proportional to body weight. The speed and jump height of the two boys would be, using the relative values just mentioned, as

$$\frac{1 \times 1}{1} : \frac{1 \cdot 12 \times 1 \cdot 25}{1 \cdot 39}$$

which is 1·40:1·39. i.e. just about 1:1. Thus, both boys in this case would have about the same dimensional pre-requisites. This also agrees with the findings in the aforementioned German studies of the speed of 11-, 12-, and 18-year old boys.

Obviously, one can draw no conclusions about motor capacity simply on the basis of a boy's or girl's age and measurements, even if a number of empirical findings suggest the existence of such a relationship. It may often be difficult to find any relationship, particularly in activities in which muscular strength plays a dominant role. Co-ordination, intelligence, and psychological make-up are other factors having a decisive influence on the shape and character of movement.

TYPES OF MOVEMENT

Just as people have been divided into constitutional types according to their external body shapes and characteristics, attempts have also been made to classify human movement. Anatomists, physiologists, psychologists, ergonomists, embryologists, sports instructors, etc. are some of the people who have sought a systematic classification of man's richly varied pattern of movement. On the basis of existing, deficient knowledge of the development of movement and control of movement from the central nervous system, it is impossible to make

9

up any adequate and informative classification. The boundaries between classes are too diffuse. Congenital and acquired movements are differentiated, and one speaks of conscious, voluntary movement in contrast to unconscious, reflex movement. But these various types are so intertwined that it is impossible to designate any one of our habitual movements as exclusively congenital, exclusively acquired or exclusively controlled by will. There are always a number of unconscious, congenital and acquired reflexes in consciously performed movements. Many of the reflex movements we make daily and incessantly were once more or less conscious movements controlled by will.

Ballistic and controlled movements are differentiated according to the manner in which the movement is produced. In ballistic movement, the musculature is only active during a brief initial phase. However, movement is controlled if both agonists and antagonists are active throughout the cycle of movement.

People often speak of natural and unnatural movements. Natural movements are those movements which generally occur in our daily lives and which may be regarded as more or less fixed movements, being made using relatively simple muscle co-ordination.

The main points in a system in which movements are classified according to their development and function and their CNS organization, will be summarized as follows. In recent years, this system has been very useful in the analysis of working movements. If a movement cycle is to be corrected in accordance with the biological and mechanical laws governing man, it is necessary to use a system, among the multitude of movements encountered, based on human anatomy and physiology.

If an attempt is made to analyse, e.g. gripping movement, chewing movement, lifting movement, balancing movement, or the human gait, one finds that three different kinds of movement are included in all cases: postural movements, locomotor transfer movements, and manipulative movements. Even if there is overlapping among the different types, this breakdown is justified for several reasons. The types of movement, or if one prefers, movement components, are functionally different. They are organized in different ways in relation to the environment, i.e. they have different stimuli and the main part of their innervation pattern lies at various levels of the CNS.

Postural movements regulate the relation of the body and body segments to gravity and acceleration. The afferent flow of impulses which primarily produce postural movements emanate from specific sensory cells in the vestibulum and semicircular canals of the inner ear, from proprioceptors in muscles, ligaments, and joint capsules, from visual receptors, and from cutaneous receptors. The discriminatory part of the innervation pattern governing postural movement is located in the medulla oblongata, mesencephalon, and vestibular part of the cerebellum, i.e. in the phylogenetically oldest part of the brain.

The afferent impulse flow to the innervation pattern for transfer movement stems mainly from those body parts which will make the movement. The body's bilateral symmetry is felt to play a decisive role in discrimination in the innervation pattern of transfer movement. If the same impulses come to discriminatory neurones from the right and left side simultaneously, no movement is produced. But if the impulses differ from one another in, e.g. time or space, the neurones react and the appropriate muscles are activated. Receptors are moved during movement and new differences arise continuously, not only differences between the right and left side but even among different sections on the same side, producing the stimulatory afferent impulse flow. Transfer movement is integrated with postural movement, which controls, *inter alia*, the extent and direction of transfer movement, and with manipulative movement. The innervation pattern of transfer movement is located in the spinal part of the cerebellum, the mesencephalon, the basal ganglia, and cortex.

Transfer movement can be extremely accurate and well-controlled as in the movements of the arms and legs in using wheels, levers, and pedals. In these cases, transfer movements are integrated with manipulative movements of the hands and feet. But they may also be free as in throwing and in swinging movements in walking and running. Manipulative movements are more subtle movements. Gripping movements by the

11

hands and fingers, movement of the jaws and lips, the foot's contact movement against the ground, and movements of the eyes are examples. These manipulative movements are adapted to the shape, position, and structure of the external object. The decision to grip an object with the whole hand or with a thumb–forefinger grip is based on visual information. Impulses from touch receptors in the fingertips provide sufficiently early information on the hardness of a piece of toast, for example, so that the appropriate biting force is applied. The cerebral part of the cerebellum, the projection system of the cerebral cortex (including the pyramidal system and its corresponding sensory system) are the CNS levels which are of fundamental importance to the innervation pattern of manipulative movement.

MOVEMENT TRAINING

In summary, one could say that man's locomotor system and locomotor patterns are organized and adapted to his environment. Environmental stimuli, such as visual impressions, touch, gravity, sound, etc., provide the main impulses for movement. In tests of locomotor function, some similar form of afference should provide the basis for the design of movements.

The generally accepted notion that practice makes perfect is especially applicable to man's locomotor function. It is known that physical activity favourably influences general metabolism and growth, and that appropriate training can increase muscular strength and endurance. However, little is known about the mechanism for development of new patterns of co-ordination. It has been experimentally determined that complete inactivity has a far-reaching effect on reflexive ability. The efficiency of the synapses declines. Thus, one may expect to find that physical activity contributes towards improved reflex co-ordination through increased synaptic efficiency. With any special form of movement or movement sequence, experience has shown that the more often the movement is made, the faster and more accurately it can be made (and the less exhausting it is). The movement's innervation pattern acquires a more fixed organization, so to speak.

## Development and Organization of Movement

The ability to perform conditional, co-ordinated movements is not congenital. It is something which must be learnt, and if a movement is to be made in the correct manner, certain characteristic conditions must be satisfied. Perhaps one sees someone execute a movement or one receives oral instructions on how to execute a movement and then tries to make that movement. Even though one's complete attention may initially be concentrated on executing the movement as it was seen or described, one finds one is unable to make the movement in the functional manner desired or anticipated. However one should really not expect to be able to perform a new, conditional movement, requiring the collaboration of several groups of muscles, merely on the basis of visual or oral information about the movement transmitted to the nervous system. The organization of a fully developed innervation pattern requires afferent impulses from proprioceptors and touch receptors, in particular from body parts taking part in the movement, i.e. the entire afferent impulse flow released when the movement is performed in its proper context.

The cerebrum, which played a vital role in the innervation pattern in learning, becomes less and less important, until its role is merely to start movement off, while subordinate centres ensure that the impulse flow to different muscles is distributed in the correct manner. Repetition of the movement is of primary importance. The more often the movement is repeated, the more automatically and functionally it is performed. Outflow to muscles facilitating movement becomes better balanced, as is also the case with inhibitory impulses to muscles antagonistic to the movement. These muscles were perhaps much too powerfully activated in the beginning, resulting in tensed muscles and heavily controlled movements. But gradually these movements become more and more fluid as the interplay between agonists and antagonists and between forces facilitating and inhibiting movement and those stabilizing the joint are perfected and become economical.

It has been estimated that the central nervous system is capable of storing more than one million innervation patterns and that locomotor learning pre-requisites are best between the ages of 8–10 years.

Muscular strength, as well as co-ordination, determines the

13

order in which muscles are activated. A well-trained person often engages antagonist muscles in order to achieve maximum performance. The effect of and ability to develop such co-ordination is to a large extent dependent upon muscular strength which, in turn, is dependent upon training, which indirectly may contribute to the design of reflex patterns and improved co-ordination.

Well-co-ordinated musculature is not always essential and decisive to muscular performance. In many instantaneous movements, muscular strength is the decisive factor. In sports such as weight-lifting and shot-putting, and in work in which heavy lifting occurs, muscular strength is what is mainly called for to execute these relatively simple patterns of co-ordination. The acquisition of increased muscular strength also requires a somewhat different method of training than in practising a co-ordination pattern.

Since muscle fibres or (more accurately) motor-units are recruited successively in a prescribed sequence, the load must be great enough to ensure that all the muscle's motor-units are engaged in muscle contractions and receive the stimuli necessary for muscle fibres to be rebuilt structurally and biochemically and adapted to increased functional requirements. Generally speaking, one could say that strength training should be undertaken under heavy loads. Opinion is divided as to how heavy these loads should be or how many contractions the muscle group will perform in every training session. This question will not be discussed here.

The strength (here we mean the mean maximum strength of a muscle or synergist group) is taken to mean the torque of the muscle or synergist group on the joint on which the muscles act. Thus, muscle strength is dependent on the tension supplied by the muscle and on the length of the muscle's lever arm. The magnitude of the tension is, in turn, dependent upon the muscle's thickness or, more properly, its physiological cross-section, the degree of muscle stretching, the number of active motor-units, and an adequate blood supply. The muscle's lever arm, like its degree of stretching, varies with the position of the joint. When the muscle is shortened in contraction and the joint position is altered, it is not uncommon for the lever arm to become longer at the same time as the shortened muscle's ability to produce tension declines. These conditions make it impossible to indicate *a priori* in which joint position muscular strength is greatest. In addition, the ability to activate the maximum number of motor-units varies in certain muscles with joint position. At the moment, there is no way of measuring muscular strength directly, i.e. the muscular torque of a separate muscle in a synergist group.

The maximum strength of a synergist group is generally measured with the aid of various kinds of dynamometers (*see* Chapter 2). Even if there are many such instruments satisfying every conceivable demand for accuracy and reliability, all measurements of muscular strength still pose major problems, particularly in the comparison of different measured values. In such studies, one is always dependent upon the subject's motivation. The ability to produce maximum force varies among different individuals and in each individual from one day to another. By simultaneous measurement of muscle power and registration of muscle activity, it is possible to some extent to check that the number of active motor-units is the same, at least approximately, at different measurement sessions. But despite this check, one must always be extremely careful when comparing muscular strength, e.g. among different persons or before and after a training period. Occasionally, one requires that the muscles will be able to make repeated contractions intermittently, perhaps for several hours, when someone runs and walks, for example, or when manual work is performed repetitively. In addition to the co-ordination and strength required to perform such a contraction, endurance, or what is often called condition, i.e. the ability to repeat contractions without exhaustion, is also demanded of the muscles. A muscle develops such condition through repeated contractions with small loads. The effect of such training, if pursued long enough, is improved circulation in the muscles as the result of increases in the capillary network. One may assume that long distance runners, who seldom display especially muscular legs, have musculature with such good condition.

GERIATRIC CHANGES

Even if no external signs of physiological changes caused by aging generally appear until advanced age, degenerative changes in the locomotor system may often appear in a person's twenties. It is usually cartilage which degenerates early. The reason for this degeneration, which consists, *inter alia*, of increased water loss, is not known. Presumably the lack of blood vessels in joint cartilage and, accordingly, impaired nutrition play a vital role.

There is much evidence to suggest that sedentary work with monotonous and fixed working positions accentuates these geriatric changes. Both clinical experience and experimental studies have shown that active joints stimulate the nutrition of cartilage and improve its elasticity while fixed and passive joints accelerate degeneration of joint cartilage. Connective tissue in joint capsules, ligaments, and tendons also change with the years. With increasing age the number of elastic fibrillae and water content in collagenous fibres decline. A consequence is that the elasticity of connective tissue also declines.

As previously mentioned, general muscular strength in man is greatest around the age of 25 and then declines slowly. It is a generally known fact that muscular strength rapidly declines in people who do not have time or opportunity to train and use their muscles. Muscular strength declines with increasing age even in people who do train or who use their muscles in hard, physical labour. However, this decline in muscular strength is not equally fast in all muscles. The strength of the back muscles and the hand's flexor muscles is reduced early. In a 37- to 38-year old, strength has declined to the level of a 20-year old. The wrist's extensor muscles and elbow's flexors retain their power somewhat longer. A 45-year old and a 20-year old can produce about the same maximum strength with these muscles. A 60-year old is about on the same level as a 20-year old, in respect of finger strength. A 65-year old man retains only half to two-thirds of his general maximum muscular strength, which corresponds to the muscular strength of a 25-year old woman. The percentage loss in muscular strength is about the same in men and women. A Japanese study of the strength, speed, and precision of movements showed that the decline in muscular strength is most apparent in the higher age groups. Precision was affected less and speed was altered least.

The skeleton also undergoes geriatric changes. However, it is the geriatric changes in joints, muscles, and the nervous system which set their mark on the movements of older people.

The result of this process is that the range of joint movements declines, movements become more deliberate, and body position more fixed.

# 2 Kinesiological Recording and Measurement Methods

A movement is often termed 'right' or 'wrong', 'suitable' or 'unsuitable', 'efficient' or 'inefficient' without any real support for such designations. In certain contexts, however, subjective evaluations of this kind are not satisfactory. In the design of places of work, position of levers and control buttons, in studies of work movements for piece-work, in the correction of movements in gymnastics or sports, or in the study of movements in orthopaedics or physiotherapy, etc., information on the nature of the movement is required which is as objective and accurate as possible. Such information is obtained by kinesiological analyses of movement.

Such an analysis may be kinematic, kinetic or both. A kinematic analysis is a geometric accounting of a movement in itself, no matter how the movement arises. However, a kinetic analysis results if an attempt is made to analyse and treat the forces and moments which produce it. A kinetic analysis is often built upon a prior kinematic analysis.

The visual image acquired of a movement's geometric shape, i.e. its kinematics, merely by observing someone performing the movement, can scarcely provide the basis for analysis of that movement. Even for the well-trained eye, the image is far too summary and subjective. Our motions are so complicated and generally occur so rapidly that the eye never has enough time to record the course of movements of body segments and the successive changes in position of the joints. The need to reproduce a movement in one way or another was also appreciated from a very early date. Some form of cine or still photography was the technique closest to hand.

The photographic method is appropriate in the study of the

development of movement. Film sequences taken of children in different age groups (e.g. 5, 7, 9, and 12 months old) sitting in a high-chair with the same toy in front of them provide a complete and objective basis for a detailed analysis of the child's gripping techniques.

At places of work, kinematic studies of movement based on films and stills have acquired considerable practical significance. The analysis of movements has been developed into elementary time systems of which MTM (method-time-measurement) and the Work Factor method are most common. A work movement is divided into sub-movements, called 'elements', and the time taken to perform each such element is calculated. Examples of elements in arm movement are 'extend', 'grip', 'move', 'release', etc. All work movements, despite their number, are based on relatively few elements. Thus, in the MTM system there are ten elements: extend (the hand), move (an object), twist (the hand), apply pressure, grip, fit, release, free, eye movements, and movements of the body, legs, and feet. Time values have been established for each of these elements through extensive studies. Since it only takes a short time to perform an element and working with decimals and strings of zeroes is cumbersome, times are not indicated in minutes and seconds but in TMU (time measurement units). The time required for a 10 cm long 'extend' movement of a certain type might be, e.g. 6·1 TMU. An MTM analysis shows step by step the elements which comprise a work movement and the standard times for each element. The total of such times provides the time required to execute the entire work operation. Thus, MTM analyses contain a kinematic description and a time determination for the movement cycle. Such analyses then provide the basis for establishing contract (piece-work) rates.

The science of the influence of force on bodies at rest or in motion is called mechanics. For kinetic analyses it is therefore necessary to know something about the laws of mechanics most often applied in kinesiology. A body's inability to change its state of rest or motion is called its inertia. The basic mechanical concept, force, is thus defined as that which changes a body's (or part thereof) state of rest or motion.

In kinesiology it is customary to distinguish between external and internal forces. Gravitation is the most important external force, as it constantly acts on the body and the body's different parts. Other external forces are air resistance and friction between the soles of the feet and the ground. Muscular forces dominate the internal forces, which are forces produced by the body itself. Friction between joint surfaces, the resistance of joint capsules and ligaments to tensile forces and the skeleton's resistance to pressure are other internal forces.

The three basic laws of mechanics were formulated by Sir Isaac Newton. The first Law, the law dealing with inertia, states: every body continues in a state of rest or uniform motion in a straight line unless it is compelled to change that state by an impressed force. Newton's second Law deals with force and shows the relationship between a body's mass, $m$, its acceleration, $a$, and a force, $F$, acting on the body. The formula for this relationship is $F = m.a$.

A body moving at a constant velocity will retain this velocity indefinitely if no external force is impressed upon it. But if a retarding force is exerted upon the body, its velocity will decline successively until the body stops altogether. The decline in velocity per unit time is great if the retarding force is great and/or the body is very light, i.e. has a small mass. The same is true if a certain force is applied to increase a body's velocity. The increase in velocity per unit time then depends on the magnitude of the force and the mass of the body. In principle, there is no difference between increase or decrease in velocity per unit time, and the general increase in velocity per unit time (per second) is called acceleration. Thus, according to Newton's Law dealing with force, it does not matter if a force acts to increase or decrease (to retard) velocity.

Using Newton's equation on force, it is thus possible to calculate acceleration, $a$, as soon as a body's mass, $m$, and the force, $F$, acting thereon, are known. Conversely, the force can be calculated which is required for mass, $m$, to perform a movement with acceleration, $a$. This latter procedure must be used when the body's internal forces are to be calculated, e.g. the force required to perform a certain movement. However, external forces acting on the body, e.g. the force between floor and foot or between hand and handle, can be measured with the aid of various methods which will be discussed later. Since rotatory movements are those which are of greatest interest in kinesiological examinations, it is practical to use another form of Newton's equation on force: the total moment of force, $M$, is equal to the body's moment of inertia, $J$, multiplied by the body's angular acceleration, $\ddot{\theta}$. The moment of force equation can be written: $M = J.\ddot{\theta}$, a formula analogous to that for the force equation, $F = m.a$.

A body's resistance to changes in its state of rest or motion is called, as already mentioned, the body's inertia. If there are rotatory movements, it is said that the body has a certain moment of inertia with respect to the axis around which the rotation occurs. By moment of inertia in this context we mean a body's resistance to rotatory movement. The magnitude of the moment of inertia depends on the body's shape and mass. For bodies of modest size in relation to the distance of the centre of mass from the axis of rotation, the moment of inertia is equal to the body's mass multiplied by the square of the distance between the axis of movement and the body's centre of mass ($J = m.r^2$). If the body is rather large, to be exact, if it is not a point-shaped object, the moment of inertia can be calculated if the body is homogenous and simple in shape. Thus, for example, the moment of inertia of a homogenous steel rod rotating around one of its own ends is equal to $\frac{1}{3}l^2.m$ where $l =$ the length of the rod, and $m =$ its mass. The moment of inertia for a cylinder rotating around its longitudinal axis is equal to $\frac{1}{2}m.r^2$ where $r =$ the radius of the cylinder and $m =$ its mass.

The moment of inertia for a certain axis in or outside a body is equal to the sum of the moments of inertia for a parallel axis through the body's centre of mass, and the product of the body's mass and the square of the internal distance between the various axes. If a body's moment of inertia around an axis through the body's centre of mass is equal to $J_\text{T}$, then the body's moment of inertia around an axis parallel to the previous axis at distance $b$ from centre of mass is $J_\text{T} + m.b^2$. It follows from this assertion that among parallel axes the axis through the centre of mass provides the smallest moment of inertia. The moment of inertia is expressed in kg.m² or g.cm².

Body segments are neither points nor homogenous bodies.

21

Nor can the centres of mass for a living person's body segments be indicated with exactness. Therefore, when the aforementioned laws governing the magnitude of the moment of inertia are used to calculate the moments of inertia of body segments, one must clearly understand that the calculated values are very approximate. In recent years, however, very painstaking examinations have been made of the distribution of mass and density in various body segments, leading to the establishment of formulae for determining local moments of inertia for body segments, i.e. moments of inertia with respect to movement axes passing through the centres of mass of the body segments. These formulae are to be found in the U.S. Air Force report no. AMRL–TDR–63–18. Once a segment's local moment of inertia has been determined, these formulae can then be used to calculate its moments of inertia around axes which do not pass through the centre of mass, e.g. the hand's or the forearm's moment of inertia with respect to the shoulder. The manner in which a body segment's moment of inertia can be determined more exactly using the quick-release method will be described later.

In the same way that forces are divided into external and internal forces, their moments of force can also be regarded as external and internal moments of force. By internal moment of force we mean the torque produced by the musculature with respect to the joints upon which the muscles act. Thus, angular acceleration and a body's moment of inertia must be determined before an internal moment of force can be calculated. We will subsequently describe how angular acceleration can be estimated or measured.

Newton's third Law asserts the equality of 'action and reaction': if a body *A* exerts a force (force of action) on body *B*, then simultaneously the body *B* exerts an equal and opposite force (force of reaction) on the body *A*.

According to the M.K.S. system, the internationally accepted system of measurement in science, force is expressed in newtons (N), velocity in metres per second (m/s), mass (weight) in kilogrammes (kg), and electrical current in amperes (A). One newton $= 1 \, \text{kg} \, \text{m/s}^2$ is the force which produces an acceleration of one metre per second per second ($1 \, \text{m/s}^2$) in a mass of one kilogramme.

In the C.G.S. system previously used, force was measured in dynes and in kilogrammes. A 1 kg force is equal to about 981,000 dynes in the C.G.S. system and 9·81 newtons in the M.K.S. system. A table is provided (Table I) which shows units and parameters in the various systems of measurement.

TABLE I

Units and parameters in various systems of measurement

| Parameters and symbols | | SI and M.K.S. system | Technical system of measurement | Unit conversions |
|---|---|---|---|---|
| Length | $l$ | 1 m | 1 cm | 1 m = 100 cm |
| Mass | $m$ | 1 kg | 1 kg | |
| Time | $t$ | 1 s | 1 s | |
| Velocity | $v$ | 1 m/s | 1 cm/s | 1 m/s = 100 cm/s |
| Acceleration | $a$ | 1 m/s² | 1 cm/s² | 1 m/s² = 100 cm/s² |
| Force | $F$ | 1 N[1] | 1 kg = 1 kp[2] | 1 N = 0·102 kp |
| Pressure | $p$ | 1 N/m² | 1 atm[5] = 1 kg/cm² | 1 atm = 98040 N/m² |
| Work | $A$ | 1 J = 1 Ws[3] | 1 kgm | 1 Ws = 0·102 kgm |
| Power | $P$ | 1 W = 1 Nm/s[4] | 1 kgm/s | 1 W = 0·102 kgm/s |
| Torque | $M$ | 1 Nm | 1 kgm | 1 Nm = 0·102 kgm |

(1) newton
(2) kilopond = the force which one kilogramme exerts on a surface under the influence of acceleration due to gravity = kilogramme force
(3) W = watt
(4) 1 Nm = 1 Ws = 0·239 gcal or 1 kWs = 0·239 kgcal 1 Ws = 1 J (joule) = $10^7$ erg
(5) atmosphere

Certain other laws have been derived as a result of Newton's Law concerning force which aid in the study of various states of force and motion. Thus, the concepts momentum, impulse, and kinetic energy must be used in certain contexts. The product of a body's mass, $m$, and its linear velocity, $v$, is called momentum. The change in momentum is called impulse, $p$. Impulse may also be written as $p = m.v_2 - m.v_1 = m(v_2 - v_1)$ when the velocity of mass, $m$, increases from $v_1$ to $v_2$. According to Newton's Law dealing with force, $m = F/a$, and since acceleration is the increase in velocity per unit time, i.e. acceleration $= dv/dt$, the impulse equation can be written in the following form:

$$p = m.(v_2 - v_1) = m \int_{t_1}^{t_2} a.dt = \int_{t_1}^{t_2} F.dt$$

23

Impulse is the time integral of force.

The mechanical work ($W$) involved in moving a body from one place to another is defined as the product of the body's resistance and the distance which the body moves according to the formula $W = F.s$. However, in the case of muscular work this definition of work is not suitable, as all static muscular work then has a zero value. It is more suitable and in closer agreement with physiological conditions to indicate work in terms of impulse measured in newton-seconds. Thus, the force expressed in newtons which a muscle produces in a given number of seconds indicates the muscle's work. According to this method of expressing work, a muscle's work may be of the same magnitude, whether the work is static or dynamic. This is something of a weakness in the expression, since a muscle's performance differs, depending on whether or not there is movement.

A body in motion, as already mentioned, possesses a certain momentum. In the study of rotatory movements, it is appropriate to introduce the concepts angular momentum and moment of impulse. A body's angular momentum, $H$, is equal to the product of the body's moment of inertia, $J$, and its rotational velocity (angular velocity, as it is commonly called), $\dot{\theta}$. The angular velocity is expressed in radians per second. A radian is $57 \cdot 3°$. The change in angular momentum is called the moment of impulse and is equal to

$$\int_{t_1}^{t_2} M.\mathrm{d}t$$

if $M$ is the moment of force and $t_2 - t_1$ is the time during which the forces and moments act.

A figure skater always begins a pirouette with outstretched arms and the feet slightly apart somewhat. After a few revolutions when the body has developed a certain angular momentum, the arms are pressed close to the sides and the feet are pressed together. Angular velocity accordingly increases sharply, since the body's moment of inertia declines as the body mass is concentrated closer to the axis of rotation. The pirouette can then continue until angular momentum has petered out. In general, the pirouette is slowed down much sooner by stretching out the arms, thereby increasing the body's moment of inertia.

A high diver is unable to change his/her angular momentum or alter the path of movement followed by the body's centre of mass once his/her feet have left the diving board. But the diver is able to alter the body's moment of inertia and, accordingly, its angular velocity. The diver leaves the board with the body almost fully extended and then has a large moment of inertia and low angular velocity. But then he/she bends forward, tucks in the legs with maximum flexion of the hips and knees, and clasps the hands in front of the lower legs, thereby reducing the moment of inertia and increasing angular velocity. This is the manoeuvre which makes it possible for him/her to rotate the body through several revolutions in the course of a drop of a few metres.

A muscle passing a joint has a certain torque, $M$, with respect to the joint's movement axis which is equal to the muscle's force, $F$, multiplied by the length of the muscle's lever, $l$, i.e. the perpendicular distance from the muscle's direction of pull to the movement axis. The magnitude of this torque is equal to the product of the rotating body's moment of inertia, $J$, and its angular acceleration, $\ddot{\theta}$. If the movement axis is not fixed, this equation of moment must be supplemented by one or more terms which take into consideration the velocity of the movement axis. Such a condition exists, for example, in the movement of the lower leg in walking. We will discuss this point later. If the magnitude of the torque around the fixed axis has been determined in this manner, the work performed, $W$, can be calculated according to the formula $W = M.\phi$ where $M$ equals moment (in newton metres) and $\phi$ equals the angle (in radians).

Figure 1 shows how it is possible to get an idea of the magnitude of the moment in static conditions.

An outstretched arm weighing 8 kg holds a 2 kg weight in its hand. In order to keep the arm in this position, a corresponding muscular moment, $M$, is required. $M$ is equal to muscular force in newtons multiplied by the muscle's lever which is 0·03 m long. The mass of the arm is 8 kg and its weight, i.e. the force with which the arm is affected by gravity, is $8.g = 8.9\cdot81$ newtons, acts through the arm's centre of mass which lies 0·25 m from the shoulder joint and provides a clock-

25

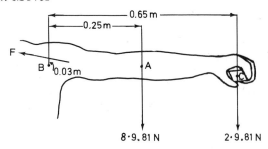

FIGURE 1. Calculation of the magnitude of a muscle moment in static conditions.

A: Arm's centre of mass; B: shoulder joint; C: weight in the hand; F: the muscle's force; $l$: the length of the muscle's lever.

The weight of the arm = 8 kg; the weight in the hand = 2 kg.

Muscular moment $M = Fl = 8 \times 9\cdot81 \times 0\cdot25 + 2 \times 9\cdot81 \times 0\cdot65$. ∴ $F = 1079$N. (*See* further in the text.)

wise torque of 8.9·81.0·25 Nm. The weight in the hand tends to press down with a force of 2.$g$ and a clockwise moment which is 2$g$.0·65 Nm when the distance between the weight and shoulder joint is 0·65 m. From the equation $x$.0·03 = 8.9·81.0·25 + 2.9·81.0·65 one obtains $x = 1079$ N.

Movement in the knee in habitual gait can be termed a pure hinge movement. In the swing phase, the lower leg performs both translation and rotation movements, Figure 2. Translation movements occur both on the horizontal and vertical planes. The forces producing these translation movements are governed by Newton's Law dealing with force, i.e. the mass of the lower leg multiplied by the acceleration of its centre of mass on the horizontal plane and vertical plane respectively. Calculation of the torque which has a rotatory effect on the lower leg is similarly based on information on the mass of the lower leg but also on the distance between the knee's axis of movement, $O$, and the centre of mass, $u$, plus the moment of inertia of the lower leg with respect to the knee's axis of movement. Since the knee axis continuously moves forward during the swing phase, the change in position of this axis in relation to a fixed co-ordinate system in space, in other words the knee's $x$ and $y$ co-ordinates, must be determined. From this information the internal or muscular moments can be calculated according to the following kinetic equation:

26

$$\text{Moment} = I\ddot{\theta} + mgl \sin\theta + ml\dot{\theta}\dot{x} \cos\theta + ml\dot{\theta}\dot{y} \sin\theta$$

$I$ is the moment of inertia of the lower limb with respect to the knee's movement axis, $l$ is the distance between this axis and the lower limb's centre of mass, $\ddot{\theta}$ is the lower limb's angular acceleration ($\dot{\theta}$ = the lower limb's angular velocity), $g$ is the acceleration due to gravity, and $m$ is the mass of the lower limb. $\theta$ is the angle formed by line $l$ with the vertical line through the knee axis, $\dot{y}$ is the horizontal velocity of the knee and $\dot{x}$ is the knee's vertical velocity. The term $mgl \sin\theta$ is gravitational torque with respect to the knee axis, $mg$ is gravity and $l \sin\theta$ is its lever. The terms $ml\dot{\theta}\dot{x} \cos\theta$ and $ml\dot{\theta}\dot{y} \sin\theta$ enter the equation because the knee axis, i.e. the axis of movement, is in motion.

FIGURE 2. Some quantities necessary for calculation of the knee's muscular moment in the swing phase of the leg in habitual gait.

O: The knee's axis of movement; u: the centre of mass of the lower leg; $l$: the distance between O and u; $\theta$: the angle between line $l$ and the vertical line through the knee axis; $a_1$ and $a_2$: accelerometers.

(*See* further in the text.)

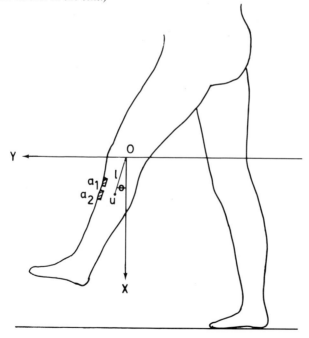

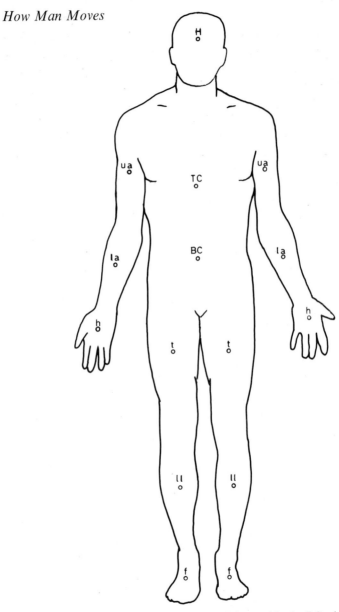

FIGURE 3. Position of the centre of gravity of the total body, BC; the centre of gravity of the trunk, head, neck, and arms, TC; the centre of gravity of the head, H; upper arm, ua; lower arm, la; hand, h; thigh, t; lower leg, ll; and foot, f.

The quantities $\dot{x}$, $\dot{y}$, $\theta$, and $\dot{\theta}$ can be measured in film sequences of the movement. If, as in Figure 2, one places on the ventral aspect of the lower limb, a pair of angular accelerometers $a_1$ and $a_2$ at different distances $r_1$ and $r_2$ from the knee axis, then the difference between the values of both accelerometers divided by the difference $r_2 - r_1$ is equal to the angular acceleration $\dot{\theta}$ of the lower limb. Studies have shown that the magnitude of the knee moment is mainly governed by the term $I\ddot{\theta}$. For practical purposes, e.g. when designing lower limb prostheses, determinations of the moment of inertia and angular acceleration are sufficient.

Mass is designated, as already mentioned, in kilogrammes and is equivalent to weight. The mass of the different body segments is expressed as a percentage of total body weight. These percentages and the position of the centres of mass of the segments may be seen in Table II and Figure 3.

TABLE II

The body segments' share of total body weight (per cent) according to Dempster's examination of eight male corpses

| Segment | Mean value |
| --- | --- |
| Head-neck | 7·9 |
| Trunk with head and neck | 56·5 |
| Right upper extremity | 4·9 |
| Left upper extremity | 4·8 |
| Right upper arm | 2·7 |
| Left upper arm | 2·6 |
| Right forearm and hand | 2·2 |
| Left forearm and hand | 2·1 |
| Right forearm | 1·5 |
| Left forearm | 1·5 |
| Right hand | 0·6 |
| Left hand | 0·6 |
| Right lower extremity | 15·7 |
| Left lower extremity | 15·7 |
| Right thigh | 9·6 |
| Left thigh | 9·7 |
| Right lower leg and foot | 5·9 |
| Left lower leg and foot | 6·0 |
| Right lower leg | 4·5 |
| Left lower leg | 4·5 |
| Right foot | 1·4 |
| Left foot | 1·4 |

In a standing symmetrical position with the arms hanging by the sides, the body's centre of gravity is in the pelvis immediately in front of the second sacral vertebra. If the body's position is changed, the position of the centre of gravity is also changed. To determine the body's centre of gravity in the standing rest position, a statograph is generally used. This apparatus functions according to the mechanical law which states that when the sum of the forces and the sum of the moments in a system of forces is equal to zero the system is in equilibrium. There are various kinds of statographs. Figure 4 shows the construction in principle. There is a disc

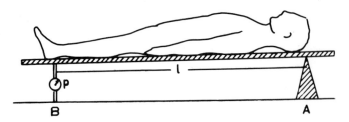

FIGURE 4. The construction in principle of a statograph.

A disc loaded with the test subject rotates around support A. Support B is a pressure gauge. The distance between A and B is *l*.

(*See* further in the text.)

on supports A and B. When the disc is loaded it rotates around the edge of support A. Support B is a pressure gauge which records the pressure the test subject on the disc exerts against support B. If the subject lies on the disc with the crown of the head (the vertex) on a level with support A and the subject's weight, $k$, is known as well as the distance between the disc's two supports, $l$, and the pressure, $p$, produced by the subject on support B, as displayed on the pressure gauge, it is possible to calculate the body's centre of gravity according to the equilibrium equation $kx - pl = 0$, i.e. $x = pl/k$.

Accordingly, the body's centre of gravity is at distance $x$ from the crown of the head. The literature generally expresses the position of the centre of gravity as a percentage of total body length reckoned from the soles of the feet. This percentage varies from 55 to 57 per cent.

Photogrammetry can be used if a determination is desired

of the centre of gravity in a certain phase of a movement or in an asymmetrical position, e.g. a forward leaning position as in Figure 5. The picture is placed as in the Figure in a rectangular co-ordinate system. The various joint axes and centres

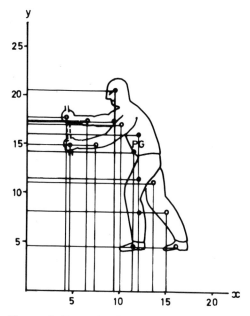

FIGURE 5. Determination of the body's centre of gravity, PG, in an asymmetrical position.

1. Calculate the product of each body segment's relative share of total body weight (*see* Table I) and the *x*-co-ordinate of its centre of mass. (The circles indicate the centres of mass of the body segments.)

2. Divide the amount of all the products by the centre body weight. That gives the *x*-co-ordinate of the point PG. Then the *y*-co-ordinate of the point PG will be calculated in a similar manner.

of mass of the various body segments are drawn in and each centre of gravity's *x* and *y* co-ordinates are measured. The torque of the body segments around the zero point of the co-ordinate system is then calculated, using the body segments' relative share of total body weight as values for 'force' and the co-ordinates as levers. The mean value of these torques provides the position of the centre of gravity. Figure 5 shows this calculation.

31

## How Man Moves

Occasionally a photo-frequency study is undertaken to provide information on positions of work. The same procedure is used as in a conventional frequency study but with the difference that a series of photographs are taken of the worker instead of noting the worker's duties at different moments. An analysis of such photographic material, which may amount to hundreds of photographs taken on random occasions, shows what causes physiologically unsuitable positions of work as well as the various positions the worker is required to adopt during his work.

FIGURE 6. Chronocyclography. *(Photo: S. Brundell)*

Lamps placed on the subject. The rotating disc in front of the lens has five long holes and one short hole.

Chronocyclography, Figure 6, is a photographic method frequently used in kinesiology. By making a series of exposures at short and known intervals on the same negative, a picture series is obtained showing the course of a movement. Exposures may be made in various ways. The camera lens may remain open while a disc rotates in front of the lens. The disc has one or more holes and a known speed of rotation. The holes pass the lens at pre-determined intervals. Another method is to illuminate the subject with periodically blinking lights from a stroboscope. Indicators are usually placed on the subject to facilitate analysis of the pictures obtained. The most suitable place for these indicators is the joint axes. The film is then projected, one frame after another, against a paper upon which the details required for analysis have been drawn. Occasionally each frame is used, occasionally every other or every fifth frame, etc., depending upon the accuracy with which the analysis is to be made.

In addition to providing a very good kinematic picture of the course of a movement, the film sequences can also be used for kinetic analyses, particularly in cases in which the movement is only made on one plane. Figure 7 shows one example.

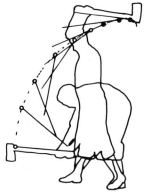

FIGURE 7. Kinetic analyses of the chopping movement with the help of chronocyclography.

The purpose here is to compare the technique of different workers in performing the chop (i.e. a chopping movement). If the path of the axe is followed, it is found that the distance between the points (the exposures) in the line of light at the beginning of the movement is very small and then increases successively the closer the axe comes to the target. Since the time between points of light is constant, the distance between these points of light is a measure of velocity. The velocity can be easily calculated when the speed of the rotating disc in front of the camera lens is known. Guided by the velocity and direction of movement in different parts of the path of movement, one can calculate the acceleration between these points and, accordingly, forces, energy, and power. Newton's equation dealing with force provides the basis for calculation of the forces which develop.

With photography of movements which proceed on more than one plane, two cameras facing one another at right angles or mirrors placed at 45° angles to the plane on which pictures are to be taken can be used. Stereo photography has also come into use in recent years. Two parallel film cameras about 60–70 cm from one another are used and exposures are made synchronously. Both films are then projected onto a screen using two projectors and are viewed through polaroid glasses, thereby producing a three-dimensional picture. Measurements of this picture can be made in the following manner. There is a reference mark on the screen. By moving the screen parallel to itself in all three dimensions, the reference point can be made to coincide with one and the same detail in both stereo images. The length of this displacement then provides the detail's change in position from one pair of pictures to the next.

In kinesiology, as in many other fields of research, a large arsenal of electronic instruments and equipment is available nowadays which can record direction of movement, movement speed, acceleration, force, etc. We did not intend to provide a comprehensive or exhaustive survey of all these aids but will describe some of the most common ones. Readers desiring more comprehensive information are referred to the appropriate technical literature.

A strain gauge usually forms part of these instruments. The strain gauge measures the change in length which occurs in a rigid body when the body is loaded. If the body's elastic properties are well known, the body's change in length can be measured electronically. The electrical measuring system is made up of a gauge which converts the rigid body's change in length to a voltage proportional to the change. The gauge may be designed according to several different principles. A common type of gauge is built around an induction system, Figure 8. Such a system consists of a coil, S, whose inductance varies when a movable vane of magnetic material, A, moves in

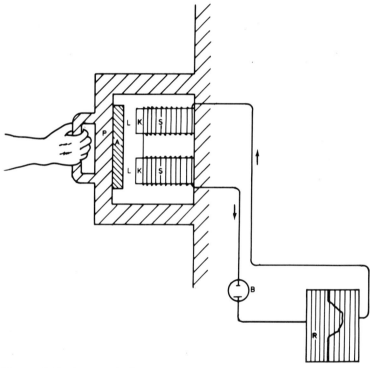

FIGURE 8. Strain gauge built around an induction system.

Coil, S; iron core, K; magnet, A, connected to the rigid body, P; battery, B; recorder, R; the air gap between core and magnet, L.

relation to the coil. The inductance in coil S is dependent upon the magnetic resistance in the circuit formed by the iron core, K, the air gap, L, and the movable vane, A. If the air gap is varied as, e.g. when the plate, P, to which the vane is attached is loaded, this means that the circuit's magnetic resistance is altered and, accordingly, the coil's inductance. The coil's inductance change becomes a measure of the vane's movement. By connecting the coil into an inductance bridge, the change in inductance can be transformed into an electrical signal.

In resistive gauges, change in the body's length is converted into a variation in electrical resistance in a thin metal wire or metal foil. In strain gauges, which have become extraordinarily important to kinesiology, the thread is embedded in an appropriately insulating material or placed between two thin layers of paper glued together, Figure 9(*a*).

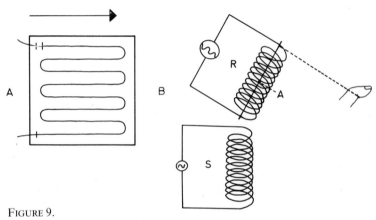

FIGURE 9.

A: A resistive strain gauge of thin metal wire.

B: An elgon consists of electrical coil systems, an external stator, S, and an internal rotor, R. The rotor rotates around axis A.

In physics, one is taught that electrical resistance $R$ (expressed in ohms) in a wire is $= cl/q$, where $l =$ the length in metres of the conductor, $q =$ its cross-sectional area in square millimetres, and $c =$ the specific resistivity. If the length of the wire and/or its cross-section are changed by stretching the wire, the wire's electrical resistance is also changed. The strain gauge is attached to its base in such a way that it accurately follows any deformation in that base. The electrical resistance varies with the magnitude of the deformation which, in turn, reflects the load to which the base is subjected. The strain-sensitive wire consists of several loops with a longitudinal direction parallel to the direction of stretching (strain). When part of the wire loop is perpendicular to the axis of strain, the gauge even acquires a certain amount of rather insignificant sensitivity to strain perpendicular to the longitudinal direction. Correction should be made for this with extremely accurate measurements of complicated conditions of tension.

A bridge circuit which can be fed alternating current or direct current is almost always used to determine a gauge's change in resistance. Since strain gauges are thermosensitive, passive temperature-compensating sensors, as similar to the gauge as possible, are connected into the circuit.

The purpose of a passive temperature-compensating sensor is to have the active gauge measure both mechanical deforma-

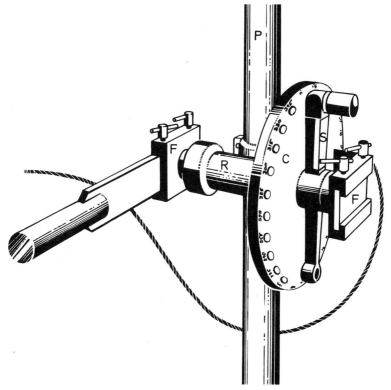

FIGURE 10. A modified Darcus dynamometer.

Pillar, P; metal tube, R, with a graduated circular disc, C, and a rod, S, attachments, F.

tion and changes in temperature while the passive sensor only registers the influence of changes in temperature. Thus, the passive sensor is put in a place which is not deformed but has the same temperature as the place at which the active gauge is sited. The difference in the changes in resistance recorded by the two becomes an expression of the mechanical deformation.

Meters with strain gauges used in kinesiology are nowadays mainly dynamometers, force-plates, and accelerometers.

Figure 10 shows a somewhat modified Darcus dynamometer. In principle, this dynamometer consists of a metal tube, R, fitted to a vertical pillar, P, attached to the floor. This tube can be raised or lowered, rotated around the pillar, and fixed in a desired position. At one end of the tube there is a graduated circular disc, C, in which holes have been drilled at 15° intervals. A rotating axle passes through the tube. Attachments, F, are affixed to both ends of the axle for different handles, pedals, or levers. A 25 cm long rod, S, is rigidly attached to the axle and the strain gauges are glued to this rod. At both ends of the rod there are holes for bolts with which the rod and, therefore, the entire axle with the handles to which force shall be applied, can be attached to the circular disc in the desired position. When someone then pulls or presses on the handle or lever, the metal rod would follow the axle's rotation if it were not fixed in place. The intended movement deforms the steel rod instead and the deformation is recorded by the strain gauges.

Other instruments in which strain gauges are used as force-sensitive detectors are the force-plates with which the forces exerted by the feet against the ground in walking and standing are recorded. A pair of such force-plates are shown in Figure 11(a). Each force-plate consists of a 400 × 400 mm lower frame and upper frame, Figure 11(b). The space inside the upper frame is filled by a robust honeycomb plate which provides the foundation for the foot. Both frames are connected to one another by force-sensitive steel shackles, four of which react to horizontal forces ($x$ and $y$ components) and four which react to vertical forces (the $z$ component). Moments around the $x$ and $y$ axes can also be recorded. Practical applications for this instrument are described in the chapter on the gait.

(a)

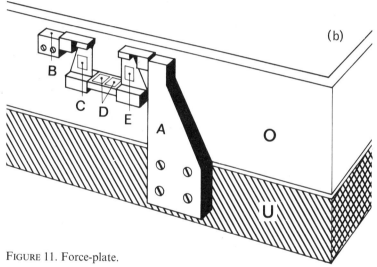

(b)

FIGURE 11. Force-plate.

(*a*) Force-plates in a walkway.

(*b*) Each force-plate consists of a lower frame, U, and an upper plate, O, connected to one another by shackles, A and B, to which strain gauges are glued (C, D, and E).

Pressure gauges are used for measurements of the heel load in shoes, Figure 12. However, these only record forces perpendicular to the pressure gauge's upper surface but make it possible to calculate the point of attack of the resultants of these forces. The instrument consists of a metal plate upon which pressure is exerted and whose centre rests on a rigid, three-armed spring balance. The support points of the spring

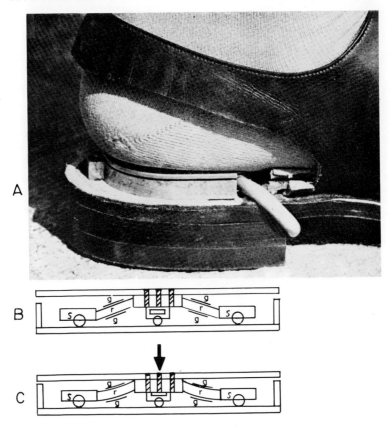

FIGURE 12.

A: Pressure gauge for measurement of heel load in shoe.
B and C: Schematic drawings of the instrument: B unloaded, C loaded.
Rigid springs, r; steel balls, s; strain gauges, g.

balance are made of steel balls which are displaced in a peripheral direction against the pressure gauge's bottom plate when the instrument is loaded. The deformation of the spring arms is measured with the aid of strain gauges. The sum of the deformation of the three arms of the spring balance provide the total heel load. A calculation of the position of the force resultants can be made from the difference in magnitude of the deformation of the individual arms. This calculation is made electronically.

To measure pressure on the sole of the foot from the sole of the shoe, small pressure-sensitive capacitors built up of sheets of tin foil and pimple rubber can be used. The gauges are part of the circuit of an oscillator. Capacity changes due to pressure on the gauge generate frequency-modulated signals.

Another method used for recording the foot's load is the piezo-electric method to which we will soon return. Since these piezo-electric elements can be made very small, a series of them can be affixed to the foot sole and thus record the distribution of pressure there.

Semi-conductor gauges are crystals (e.g. quartz or tourmaline) which produce a voltage on mechanical loading or change

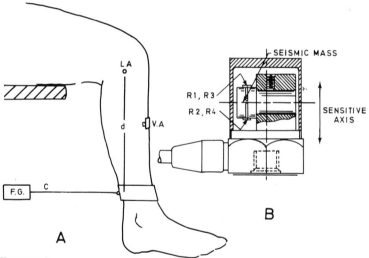

FIGURE 13.

A. Measurement of the moment of inertia for the lower leg by the quick-release method.

LA: The knee's movement axis; VA: strain gauge accelerometer; C: cord; FG: force gauge; d: distance between LA and the cord's attachment to the lower leg.

B. Sketch of a piezo-resistive strain gauge accelerometer.

The piezo-resistive strain gauge elements are identified as $R_1$, $R_2$, $R_3$, and $R_4$. These strain gauge elements are fastened on each side of slots machined in a cylindrical member. When upward motion is applied to the base of the accelerometer along the sensitive axis the mass element portion of the cylinder bends very slightly toward the base causing the length and resistance of $R_1$ and $R_3$ to increase while the length in resistance $R_2$ and $R_4$ decreases.

41

their electrical resistance in proportion to the force to which they are exposed. These gauges are also connected to a bridge and are used, for example, in accelerometers. Figure 13 shows an accelerometer for the registration of changes in the velocity of the lower leg.

Newton's Law dealing with force, $F = m.a$, provides the basis for construction of the most common accelerometers used in kinesiology. When the accelerometer comes into motion, the crystal is affected by the inertia of the weight to which it is attached, with a force, $F$ (pressure, tensile or shearing force), thereby producing a voltage in the crystal proportional to the acceleration, $a$. Thus, the crystal itself generates the voltage. In contrast to these piezo-electric accelerometers there are also piezo-resistive accelerometers. In the latter, a known voltage is fed through the crystal. When the crystal's electrical resistance changes in proportion to the force to which the crystal is subjected in the accelerometer's movement, the voltage is also changed with the degree of acceleration.

Because of the wide area of use for strain gauges, there is no clear-cut procedure or method for the choice of a gauge which is appropriate in all contexts. Experience must determine the choice among the many hundreds of different gauges available. Factors such as the object to be measured, accuracy desired, dimensions, costs, etc. decide the matter.

Accelerometers have proved to be indispensable in the accurate calculation of the moments of inertia of different parts of the body. They are used in the quick-release method. This method is based on Newton's moment of force equation for the rotary movement of bodies around a fixed axis. The equation states: $M = J.\ddot{\theta}$, where $M$ is the total moment of force with respect to the fixed joint axis and consists of the moment of gravity and the moment of muscular force; $J$ is the total moment of inertia consisting of the moment of inertia of the body segment $J_k$ and the moment of inertia $J_m$ of the gauge's attachment device to the body; and $\ddot{\theta}$ is the angular acceleration of the part of the body.

The measurement is illustrated in Figure 13 in which the moments of inertia for the lower leg and foot are to be established. A cord is attached immediately above the ankle. The other end of the cord is attached to a force gauge which indicates the cord's state of tension. The thigh and, accordingly, the knee's movement axis, LA, is fixed with a suitable device and VA is the angular accelerometer attached to the lower leg.

Subjects are requested to extend their knees and they then produce an isometric contraction of the quadriceps. If the cord's attachment is suddenly released without the knowledge of the subject, the lower leg will rotate forwards with the accelerometer, thanks to the extremely active quadriceps. The moments around LA which act at the instant the cord is loosened are produced by the musculature and gravitation. The latter's moment is, however, zero when gravitational force is immediately under the joint axis, LA. The moment of the musculature, $M$, can then be calculated quite simply. If the cord's tension, as registered by the force gauge, is equal to $F$, and the distance between LA and the cord's attachment to the lower leg is equal to $d$, then $M = F.d$. The subject simply does not have time to relax the muscles at the instant of release. This can be confirmed with electromyography. Thus, the total moment of inertia, $J = F.d/\ddot{\theta}$, contains the moments of inertia of both the lower leg and foot, $J_k$, as well as $J_m$ which can be determined for each measuring device. $J_m$ consists of, e.g., the angular accelerometer's moment of inertia and the cord's moment of inertia plus possible moments of inertia of other fixation devices with respect to LA.

$$J_k = \frac{F.d}{\ddot{\theta}} - J_m$$

For the electronic measurements of movements, particularly in separate joints, elgons (resolvers) have come into use in recent years. These units are constructed according to the principle of induction. In its simplest form, a resolver consists of electrical coil systems, an external stator and an internal revolving rotor, Figure 9(*b*). An alternating current is fed to the stator around which a magnetic field develops which, in turn, induces voltages in the rotor. The magnitude of these voltages varies with the angle formed by the rotor and stator. If the rotor is perpendicular to the stator, i.e. parallel to the stator's magnetic field, no voltages arise in the rotor. The more the rotor's long axis approaches the stator's, the greater the

43

induction voltage. These voltages are recorded on some simple recorder. The rotor axle is connected to the body part whose movement is to be recorded. This connection must be so devised that the rotor's position accurately describes the movement of the body part.

Since only one direction of motion can be measured with a resolver, three resolvers are, thus, required to record the three spatial co-ordinates. All three resolvers must be connected to the same point on the body part in question. The recorder then continuously records the movement of this point in the three spatial co-ordinates with three separate traces. The movements of the body part can be marked in a co-ordinate system. This method is especially useful for subtle movements in smaller joints.

The electrogoniographic method can be used when one wishes to measure the movement of a large joint and the direction, velocity, and acceleration of this movement. The instrument is a modified release potentiometer. Figure 14

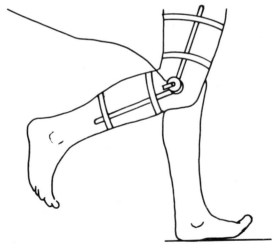

FIGURE 14. Electrogoniography for registration of movements in the knee.

shows an electrogoniograph attached to the leg for registration of movements in the knee. Figure 15 shows the movements recorded in an examination of the antagonist relationship between the vastus lateralis and the biceps femoris.

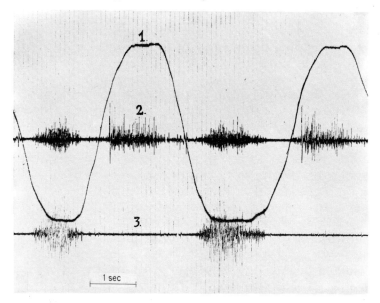

FIGURE 15.

1 : Electrogoniographic registration of extension (falling curve) and flexion (rising curve) in the knee.
2: EMG of biceps femoris.
3: EMG of vastus lateralis.

Electromyography (the abbreviation EMG will here mean electromyography, electromyographic, electromyogram, etc.) is an efficient method for recording the presence of activity in muscle.

When the function of the muscles is discussed in anatomy books, the mechanical function is usually meant, i.e. the movements the muscles are mechanically able to produce through their course in relation to the joint axes. One assumes that there is 100 per cent functional co-ordination. With flexion one simply assumes that all muscles which are mechanically able to perform flexion actively collaborate in the movement and, at the same time, that their mechanical antagonists remain passive. This way of looking at things is the basis for anatomical designations such as 'flexor', 'extensor', etc.

However, we know from electromyographic studies that the active engagement of the muscles does not always agree with

45

their mechanical ability to produce movement. Thus, it is rather uncommon to find all the muscles in a synergist group active at the same time, and a muscle antagonistic to the movement is often found to be active. A tabulation of the muscle's degree of activity as registered by electromyography and the muscle's direction of pull in relation to one or more joint axes passed by the muscles provides us with information on the muscle's true function in different movements and positions.

Like the nerve cell, the muscle cell, in the state of rest, is able to maintain an electrical potential between the cell's inner cytoplasm and the extra-cellular space. The different intra- and extra-cellular ionic composition produces this potential over the cell membrane (the membrane potential). When a muscle cell is stimulated by transfer of a nerve impulse via the motor end-plate from the nerve fibre to the muscle cell, the membrane potential is reduced, the cell membrane is depolarized, and migration of ions between the cell's interior and external environment occurs. The migration of ions through the cell membrane not only affects electrical potential but also produces a migration of ions in the extra-cellular space. These extra-cellular voltage potentials are the ones recorded by EMG.

There are thousands of muscle fibres in each muscle and these muscle fibres are arranged in functional units (motor-units), Figure 16. A motor nerve cell and all the muscle fibres innervated by this cell form one motor-unit. The number of muscle fibres in a motor-unit varies from muscle to muscle. Each unit in the outer eye muscles consists of 8–10 muscle fibres while there are motor-units in the large muscles of the leg which contain about 2,000 muscle fibres. The muscle fibres which belong to the same motor-unit react as a whole. They are activated at the same time since the same nerve impulse stimulates them. The sum of the changes in extra-cellular voltage for these muscle fibres results in motor-unit potentials (action potentials). In a relaxed muscle there are no action potentials. If action potentials do occur, this is a sign that the muscles have been activated, and the more a muscle is contracted, the greater the number of motor-units which are active and the greater the frequency of the action potentials of individual motor-units. Frequencies of up to 50 per second have

46

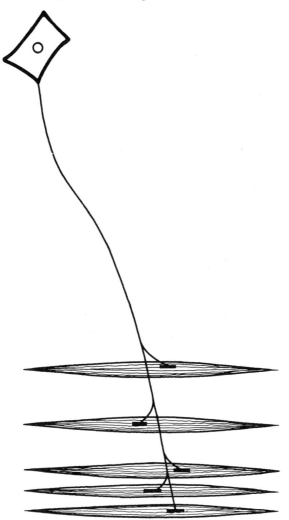

FIGURE 16. A motor-unit.

been registered in some motor-units undergoing powerful contraction. The different, active motor-units are distinguished by action potentials with different frequencies and amplitudes. With a powerful contraction, however, the different potentials cannot be distinguished from one another in EMG; an interference pattern, Figure 17, develops instead.

47

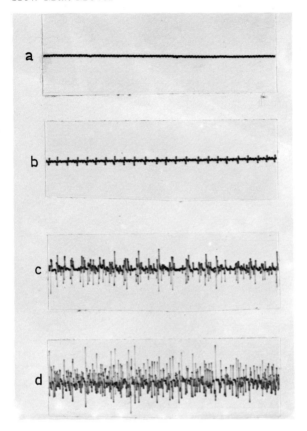

FIGURE 17. EMG of extensor digitorum.

(*a*) No activity.
(*b*) One active motor-unit.
(*c*) Slight activity.
(*d*) Marked activity.
(*c*) and (*d*) show interference patterns.

Even if an interference pattern makes it impossible to count the number of potentials or distinguish the various motor-units from one another, muscle contraction can still be graduated through a graphic integration of the EMG. The area of the myogram covered by the recorded potentials per unit time is proportional to the activity of the muscle. In addition to this graphic calculation of the muscle's degree of electrical activity in an EMG, electronic integrators can also be used. These consist of a resistor and a capacitor (RC filter) or a resistor, capacitor and an induction coil (RCL filter). The potentials are fed into such a filter and thereafter to a recorder which continuously records the numerical voltage (or more accurately, the mean voltage) of the potentials, Figure 18.

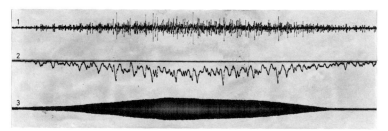

FIGURE 18. Isometric contraction.
  1: EMG.
  2: Mean voltage.
  3: Muscle force.

Even if a muscle's degree of electrical activity and, accordingly, its degree of contraction, can be graduated with EMG, no fixed conclusions can be drawn about the muscle's mechanical power on the basis of these registrations. The muscle also contains elastic, passive forces which are not reflected in muscle activity. In isometric contraction when the elastic forces are constant, the voluntary power of contraction of the muscles does increase with the number of active motor-units and with the discharge frequency of individual motor-units, Figure 18. If muscle length is altered, however, conditions become more complicated. The muscle's ability to develop mechanical power is not only dependent upon the degree of activity but also upon the muscle's degree of stretching and the load to which the muscle is subjected. The greatest power is produced by muscle fibres when they have a length equal to the length they adopt at rest. But a large part of the power lost by muscle fibres when they are stretched beyond the rest length is compensated for by the elastic forces in the muscle.

The relationship between the muscle's electrical response and mechanical response becomes even more complicated with shortening: the faster the shortening, the greater the muscular activity. As the muscle shortens, active motor-units successively take over the load which was initially borne by passive fibres. This leads to a successive decline in the speed of shortening as the shortening progresses.

Activity increases with fatigue, i.e. with exertion of the same exterior force, and the 'fatigued' muscle displays a greater degree of activity than a rested muscle; many motor-units must be engaged.

If the load is constant with isotonic contraction, there is a linear increase in the degree of activity in step with the speed of shortening. But with eccentric contraction, e.g. when someone gives way under a load, the degree of activity is not affected by the speed of movement. Only the magnitude of the load (the retarding force) then determines the degree of activity.

The accuracy with which action potentials can be registered depends on the impedance of the recording electrodes, the amplifier's input impedance and frequency range, and the ability to suppress similar polarities, i.e. two simultaneous positive potentials.

The electrical activity of the muscles, e.g. the action potentials, can be picked up by surface electrodes or by intra-muscular electrodes, Figure 19. Surface electrodes are attached to the skin above the section of muscle to be studied. Both surface electrodes and intra-muscular electrodes may be bipolar or unipolar. In bipolar recordings with surface electrodes, both the electrodes are usually identical in shape and are usually made of small plates of silver or tin. It is the potential between the two electrodes which is measured. With unipolar registration, one electrode is different and placed above the muscle in question while the other electrode is placed at a certain distance from the muscle, at a place on the body where there is little or no electrical activity, e.g. the outer ear. If activity should occur in the vicinity of the other electrode, there is a risk of this activity being picked up and recorded at the same time as the activity picked up by the different electrode, i.e. the activity one wished to record. With intra-muscular unipolar recording, the active electrode is placed in the muscle while the other electrode is usually attached to the skin.

Coaxial needle electrodes are often used in the study of

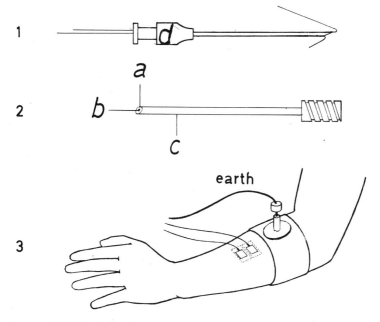

FIGURE 19. EMG electrodes.

1: Intra-muscular wire electrode.
2: Coaxial needle electrode.
3: Surface electrodes.
Insulation, a, between the platinum wire, b, and canula, c; hypodermic needle, d.

isometric muscular contractions. Such an electrode consists of an insulated platinum wire inside a canula. The tip of the platinum wire without insulation comprises one pole and the canula the other pole. Using such an electrode it becomes possible to study individual action potentials and the activity within a very small area.

However, since coaxial needle electrodes are painful and uncomfortable with isotonic muscular contractions, these electrodes are used less frequently in motion studies. In kinesiology, intra-muscular wire electrodes are used instead.

These electrodes are made of thin flexible metal wire, entirely insulated except for a varying distance at either end. The wires are usually polyurethane-coated Karma wire with a diameter of 0·05 or 0·025 mm. Usually the insulation is

51

removed from the intra-muscular end of the wire for a distance of 1–5 mm. The insulation can be removed chemically with a paint-remover, by burning in a flame, or mechanically with a knife.

The wires are introduced in pairs or separately into the muscle using a hypodermic needle. The ends of the wires are usually bent over the rim of the insertion, forming a barb which hooks the wires to the muscle tissue when the insertion needle is withdrawn.

Wire electrodes are the type of intra-muscular EMG electrodes which cause the least discomfort to subjects. Once the electrodes have been inserted they usually cause little or no discomfort. Often the subjects do not even feel the presence of the wires unless strong contractions are performed.

Radiological studies have disclosed that EMG-recording wire ends usually remain at approximately the same intra-muscular location during the whole experiment. The barbs are efficient enough to anchor the wires to the muscle even when the wires are introduced through one muscle into a deeper muscle. On the other hand, minor dislocation of the wire ends were shown to be common. These minor dislocations may cause alterations in the recorded EMG activity, which may be erroneously interpreted as a change in the degree of contraction. This is an important source of error when quantitative studies of EMG activity are performed.

Surface electrodes can often be used in kinesiology, particularly if one is interested in the activity of an entire muscle group. If, for example, a pair of surface electrodes are placed on the skin above the calf musculature, one cannot know from which muscle or muscles the recorded activity emanates, only that there is activity in the calf musculature (e.g. in plantar flexion of the foot). An electromyographic study can often be begun with surface electrodes in order to study, e.g., the co-ordination between two antagonistic groups. If a detailed study of the activity of individual muscles within the muscle group is subsequently desired, more selective electrodes such as wire electrodes must be used.

The potentials are either transmitted through cables or by telemetry to the amplifier and recording instrument. Recording is usually done with a mirror galvanometer recorder; an ink jet recorder can be used in certain cases. Electromyographs of high quality are available on the market as standard equipment. Anyone who builds his own equipment or buys it ready-made should remember that the range of voltages in EMG usually lies between 10 $\mu$V and 3–4 mV, and the frequency range from 25 to 20,000 Hz.

# 3 Mechanics of the Standing Rest Position

Technically speaking, man's upright posture may appear to be a less than successful construction. The body's centre of gravity is high above a relatively insignificant support area. In addition, a human body does not consist of a uniformly rigid, solid block but of a series of segments arrayed on top of one another and connected by movable synovial, cartilaginous, and fibrous joints. Notwithstanding, man stands rather steadily on two feet, and great force is required to throw the body off balance.

Standing rest position will here be taken to mean an upright, standing symmetrical rest position with the body weight evenly distributed over the right and left feet, the eyes straight ahead and the arms hanging limply. This is, of course, only one of the many upright postures adopted by man. An English study in 1953 showed that 1,423 out of 1,710 people unconsciously stood with the body weight on one leg or the other. They shifted continuously between right and left leg and only stood for brief periods with an evenly distributed load. Curiously enough, people seldom take the opportunity to lean against something or search for some support.

From a strictly mechanical point of view it is incorrect to regard the standing rest position as a static problem. People never stand completely still, even if they try to and believe that they are doing so. The body continuously performs postural sways in the sagittal and frontal planes. [*Note*: A photograph of the planes of the body is given with the Glossary.] These postural movements or sways are too small to be detected by the naked eye and do not prevent static measurements and mechanical analyses.

If one places an accelerometer, e.g., on the tibia at the front of the lower leg or on the forehead, one finds that the body performs two kinds of wave movements in the sagittal plane, one long-wave and one short-wave movement. The former movement is a sway of the body forward and back, essentially around the transverse axis of the ankle joint. This causes the head to describe much greater movements than the trunk and lower limbs. As a rule 6 to 8 such sways per minute are performed in which the head's movement amplitude is about 4 cm. There are large individual variations, and older people often display larger postural movements than young people. The short-wave movement is superimposed on these sways at a rate of 8–10 cycles per second. Coronal postural sways are much smaller than sagittal sways and are without importance to the remainder of our discussion. On the other hand, the magnitude of sagittal sways plays a vital role in any discussion of the interplay of forces around the various transverse movement axes.

### THE FOOT AS AN ORGAN OF SUPPORT

From the point of view of comparative anatomy, the human foot may be regarded as a climbing foot rebuilt into a supporting foot. The basic position of the climbing foot is one of supination. When the supinated foot rests on a horizontal surface, the outer margin of the foot is the supporting area, the metatarsals vertically overlie one another and the inner foot margin is turned upwards. This position is preserved in man in the tarsus, while the metatarsals and toes are heavily pronated. All of this part of the foot skeleton is twisted around the foot's longitudinal axis so that the heads of the metatarsals and toes are almost horizontal or form a mild, dorsally convex arch.

In the root of the foot, the talus is to be found above the lateral heel bone, the calcaneus. In the distal part of the root of the foot, the medial bones, the cuneiforms, are higher than the cuboid at the foot's outer margin. The proximal ends of the metatarsals, which are adjacent to and lead in towards the root of the foot, also display minor supination.

Body weight is transferred at the talus to the resilient arch formed by the root of the foot and metatarsals. This arch construction has many advantages compared to a flat support organ. With the arched foot, load is distributed more evenly on the entire foot skeleton, and the foot can be adapted to different load conditions and surfaces. When the arch construction of the foot is discussed, distinctions are usually made between the longitudinal and transverse arch. The longitudinal arch extends between the posterior surface of the calcaneum and the heads of the metatarsal bones.

The transverse arch is most bowed in the anterior part of the root of the foot where it is formed by the cuneiforms and cuboid. Proximal parts of the metatarsals also form a well-defined arch which is, however, obliterated distally.

The foot's arch construction, like distribution of pressure and foot direction to a certain extent, is reflected in the foot's internal structure, Figure 20. A direct continuation of the arrangement of trabeculae with the distal end of the tibia is found in the talus where they stream together into the body

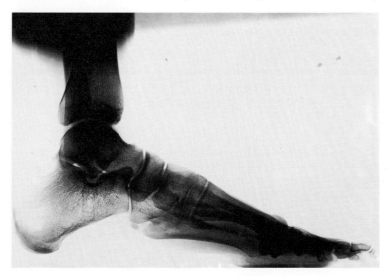

FIGURE 20. X-ray of the foot. Note the arrangement of the bone beam system.

of the talus. From this point, a beam system arises in a direction backward and downward to the posterior surface of the calcaneum and in a direction forward and down to the heads of the metatarsal bones. Thus, the arrangement of the power-absorbing bone beams can be followed continuously through the various bones, despite intervening joints.

With the exception of the talus, the tarsal bones and meta-tarsal bones are joined together in what is called the foot's fixed groundwork. Only minor displacements and movements occur between skeletal parts herein, and it is the groundwork which forms the different arch constructions.

A question often discussed is that which concerns the behaviour of the arch of the foot and of the foot skeleton during weight-bearing of the foot. The answers to this question which can be found in the literature are by no means uniform. In one 1949 textbook on anatomy it is stated, for example, that the navicular bone is depressed on weight-bearing of the foot by 6·5 mm on the average. Another author put forward 8 mm in this respect and added that the bones of the whole foot are depressed but that their mutual positions do not change when the foot bears weight. According to other investigations the foot arch is elongated 19 mm within the second metatarsal ray and 8 mm within the fifth metatarsal ray upon weight-bearing. At the same time, the foot becomes broader so that the distance between the bases of the first and fifth metatarsals increases by a mean figure of 8·5 mm. In a radiological study of the foot's change of form over a short and long period of weight-bearing, one author felt that he had been able to demonstrate that X-rays taken of a weight-bearing and non-weight-bearing foot were identical or almost identical.

In a radiological and anthropometric study of a series of clinically normal feet, two Swedish workers found (1968) that one cannot speak of any true changes in the form of the foot skeleton during momentary weight-bearing of the foot. The increase in width within the region of the ball of the foot, i.e. within the region of the metatarsal heads, which arises on weight-bearing of the foot seems, therefore, to be a soft tissue change. When the foot bears weight, the subcutaneous tissue is pressed together and displacements in a medial and lateral direction occur, producing stretching of the tissues, so that measurable increases in volume arise.

The fact that the arch of a stable and sufficient foot is not

57

C

flattened when the foot is loaded depends on several factors working together: the shape and design of skeletal parts, the powerful plantar ligaments, the plantar aponeurosis and the musculature. Any discussion of the question of which structure is the most essential would be pointless here, even if one is tempted to emphasize the importance of passive forces, primarily by the ligaments. No activity can be demonstrated electromyographically in the plantar muscles in a symmetrical standing position. This does not, however, prevent passive muscle forces from playing a certain role in the fixation of the foot's bones. A sufficient foot still undergoes a certain amount of change in shape when it is loaded. But this concerns the external shape. The foot skeleton sinks into plantar fatty tissue which is then pressed out, producing a change in the foot's external contours.

If the skeletal arch is not to be flattened and become depressed when the foot is loaded, a vital pre-requisite is that the arch's supporting points do not glide apart but are able to resist horizontal forces which arise when the arch is loaded, Figure 21. An idea of the magnitude of the horizontal force, $F$, to which the longitudinal arch is exposed in the standing symmetrical rest position, is given by the following formula: $F = Wab/Sh$.

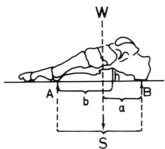

FIGURE 21. Factors for calculation of the horizontal force to which the arch of the foot is exposed.
  Supporting points, A and B; body's line of weight, W; maximum arch height, h.
  (*See* further in the text.)

Here $W$ is body weight, $S$ the distance between the arch's anterior, A, and posterior, B, supporting points, $a$ the distance between the posterior supporting point and the point at which the body's line of weight intersects the supporting area, *b* the distance between the anterior supporting point and a vertical line through the highest point of the arch, and *h* the maximum arch height. Strain on the arch's supporting capacity becomes greater as body weight increases, the more forward the line of gravity passes the supporting area (the more a person leans forward) and the lower the foot arch.

TARSAL JOINTS

Body weight is transmitted via the talus to the foot's fixed groundworks. The talus lies like a powerful bone disc between the tibia and fibula on the one hand and the calcaneus and the navicular bone in the foot's fixed groundwork on the other hand. How are the joints on each side of the talus, i.e. the talo-crural joint, subtalus, and talocalcaneo-navicular joints respectively stabilized?

A series of studies have shown that the common centre of gravity for the total body mass in the upright standing position is on a level with and immediately in front of the second sacral vertebra. This more or less means that the centre of gravity is at a distance equal to 55–57 per cent of total body length, calculated from the soles of the feet. The height of the centre of gravity is often given according to the formula $Y = 0.557X + 1.4$ cm, $Y$ being the distance between the sole of the foot and the centre of gravity and $X$ the body length.

The body's centre of gravity has, of course, no anatomically fixed position when movements are made with the arms and legs. Even when a person is apparently standing still, e.g. in a symmetrical rest position, his body's centre of gravity moves continuously because of postural sways and the movements of heart and lungs. However, these shifts in centre of gravity are small and take place within a very limited area. Thus, a vertical line through the centre of gravity, the body's line of gravity, also has a momentary position, but these changes in position are not great enough to prevent determination of the positional relationship of the line of gravity to the movement axes of the major joints and therefore the effect of body weight on these joints. Thus, it is known that the body's line of gravity

59

runs almost through the middle of the body's support area. The body's support area is the area limited by the outer margins of both feet and the lines of connection between the toe tips and heels of both feet. Seen from the side, the aforementioned vertical line is on a plane which passes roughly between both support points of the foot's long arch, i.e. right between the heel and the heads of the metatarsal bones, Figure 22. The line of gravity passes 2–5 cm ahead of the movement axis of the talo-crural joint, which itself passes through the lower tip of the lateral malleolus. Even with powerful postural sways, the vertical line still remains ventral to the joint axis. Thus, body weight acts with a forward falling force on the ankle.

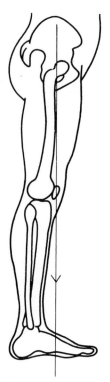

FIGURE 22. The body's vertical line passes 2–5 cm anterior to the ankle.

The ankle joint is a stable single axis hinge joint under ligamentous and skeletal control. The cylindrical upper part of the body of the talus, the trochlear surface, has a slightly curved radius which causes major reaction at the proximal end of the lower leg to relatively minor movement in the joint. The amplitude of postural sways is also much greater at the proximal end of the lower leg than at its distal end.

A sagittal cross-section through the joint shows that the joint gap is tilted backwards and down somewhat when the foot is horizontal and the tibia is vertical. From the static point of view it would, therefore, be advantageous if the lower leg in a standing rest position could lean forward somewhat. The pressure which the' lower leg exerts upon the talus could then be exerted as a force perpendicular to the surface of contact and no moment would have to arise in respect of the talus. According to Danish studies, the lower leg does in fact lean forward about 5° in relation to the line of gravity when a person stands in a symmetrical rest position.

The posterior calf muscles must intervene in order to counter-balance the body weight's forward falling moment. Electromyographic studies have shown that it is the soleus muscle which joins in to balance the body weight moment around the ankle. Muscles which act to maintain body position are usually called postural muscles. Thus, the soleus is a postural muscle.

Less frequently, one finds that the gastrocnemius or the deep flexor muscles are in action in the standing rest position. This is also true of the fibularis muscles. Nor do the anterior leg muscles, the extensors, actively collaborate in body balance in this position. This concentration of load on one single muscle is probably the most important reason why we often have to 'change foot' when standing, even though only a small part of the muscle's capacity is put into use.

THE KNEE

The joint socket in the knee is made up of the gently concave and slightly backward leaning joint surfaces on the proximal end of the tibia. However, since the lower leg leans forward slightly in the standing rest position, the joint socket comes to adopt a horizontal position and provides a broad platform as support for the femur and both its condyles. The femur's mechanical axes, i.e. the straight line connecting the centres of

61

the femoral head and condyles, and the longitudinal axis of the tibia form a straight line on the frontal plane. On the other hand, the longitudinal axis of the femur and tibia form a laterally open abductive angle of about 175°. Projected in the sagittal plane, Figure 23, the mechanical axes of the tibia and femur tilt forward and up a little.

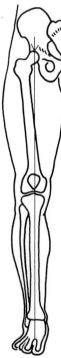

FIGURE 23. Mechanical axes of the femur and tibia.

Thus, the body's line of gravity acquires another positional relationship to the knee's transverse movement axis and to the movement axis of the talo-crural joint. In fact, this vertical line passes through the patella when a person stands in an upright symmetrical rest position.

However, the body's line of gravity is not really the subject of greatest interest in any discussion of the knee's stabilization, as the knee does not bear the entire body weight but only that part of the body mass which lies above the knee. Since the lower leg leans forward and up to 5° and the mass of the lower leg lies behind the body's line of gravity, the vertical line for body mass above the knee must pass in front of the knee's transverse movement axis. Accordingly, body weight has an extensive effect on the knee. This extensive effect must be counter-balanced in one way or another. Thus one would not expect to find the powerful knee extensor, the quadriceps femoris, active in a standing rest position. Palpation of the patella also confirms that the muscle is relaxed. The patella can easily be displaced to the side. However, if the muscle is contracted the patella is firmly fixed to the underlying femoral joint surface and the suprapatellar bursa. Electromyography has also shown that the muscle is inactive in this position.

On the other hand, one should expect to find one or more of the knee's flexors in action as a counter-force to the extensive effect of body weight. In a freely dangling leg, there are in fact no fewer than 8 muscles which are mechanically qualfied to flex the knee: the semitendinosus, semimembranosis, biceps femoris, sartorius, gracilis, gastrocnemius, plantaris, and popliteus. But in the standing position when the foot is fixed on the ground, mechanical pre-requisites are changed for most of these muscles.

The reason for this is that two-joint muscles come to act within a close muscular chain. Two-joint muscles are here taken to mean muscles which pass an adjacent joint, in addition to the knee, which is either the hip or talo-crural joint. A closed muscular chain exists as soon as body segments to which a two-joint muscle is attached remain fixed or if their movements are controlled by forces other than those of the two-joint muscle.

Figure 24 shows in greater detail the mechanical qualifications the various knee flexors have to influence movements in the knee with an open and closed chain. This mechanical analysis shows that if the gastrocnemius or hamstring muscles, for example, were contracted and included among the postural muscles, they would work synergistically with body weight and accentuate knee extension.

Electromyographic studies have shown that in most people these muscles are also inactively engaged in the symmetrical rest position. None of the other muscles passing the knee are usually active either. In certain cases, there is intermittent

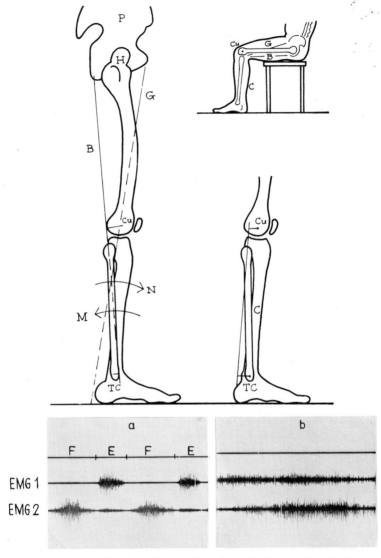

FIGURE 24. In the sitting position with the lower limb dangling, the biceps femoris, B, and gracilis, G, act as knee flexors. In the standing position when the pelvis, P, as well as the foot are fixed, the biceps will act as a knee extensor. If the tensile direction of the muscles is extended downwards, one finds that B's direction of power lies anterior to the axis, TC, of the ankle. B will therefore produce a moment around TC in the direction of the arrow, M, i.e. a back-

ward movement of the lower limb, C, which is the same as extension in the knee, and an upward lift of the entire body mass above the knee. This extensive effect is seen more clearly when someone rises from a bent knee position, EMG 2: b. However, the direction of power for muscle G passes posterior to axis TC and, therefore, produces a moment around TC in the direction of arrow, N, i.e. the muscle flexes the knee. With a fixed foot the diarthric muscles, the gastrocnemius and plantaris, have the same mechanical effect on movements in the knee as the biceps femoris. This also applies to the semimembranosus and semitendinosus. The sartorius, like the gracilis, also acts as a flexor of the knee, even in a standing position.

EMG 1: vastus lateralis, EMG 2: biceps femoris.

a: flexion and extension in the knee with dangling lower limb; b: rising from a bent knee position.

E = extension, F = flexion.

activity in the biceps femoris, but this activity belongs to the muscle's extensive action on the hip and contributes in no way to counteracting the body weight's extensive effect on the knee.

However, in the standing rest position the knee, despite body weight's extensive effect on the joint, is not extended to a maximum. The posterior capsule wall and collateral ligaments appear to be the structures which primarily counterbalance body weight and contribute towards knee fixation. The large contact area between the condyles, menisci, socket and iliotibial tract are factors which contribute to stability. Thus, no active muscle collaboration is necessary for stabilization of the knee.

However, this passive joint stabilization can only occur within a very limited area of the knee's range of motion. If a person adopts a 'stand-at-attention' posture, the knee is subjected to maximum or nearly maximum extension, and both the quadriceps femoris and the hamstring muscles are activated. They are also activated if a person does a minor knee bend. The line of gravity will then pass behind the movement axis, and body weight then acts as a flexor effect on the knee. In order to prevent further bending of the knees, knee extensors go into action, i.e. the quadriceps and the hamstring muscles.

THE HIP

The human body's rise from a previous, possibly four-footed

65

stage of development occurred around the hip. The body's centre of gravity lies accordingly above the hip. Powerful muscles have developed around the hip in order to balance the trunk on the femoral heads in different situations. Man is the only animal with a centre of gravity above the hips, and only man has hip extensor musculature powerful enough to be called a 'bottom'. The body's most powerful ligaments, the iliofemoral, are also in the hip joint, and the body's most powerful muscle, the gluteus maximus, also acts on the hip. It is no coincidence that the most powerful fasciae, i.e. the fascia of the abdominal walls, fascia lata and the lumbar fascia, have insertions at the pelvis and, accordingly, near the hip. Nor is it any coincidence that the spine is intimately connected by a joint with a very minimal amount of movement to the pelvis to form a pelvic ring.

The hip is a 'ball and socket' joint, but its range of motion is greatly restricted so as to favour stability. A number of safety devices restrict the joint's motions, devices such as a deep joint socket with a labrum, a powerful capsule and a well-developed ligamentous apparatus.

At present no exhaustive answer can be provided to the question of the hip's fixation in the standing rest position. This is mainly because it is still not known whether or not the deepest hip muscles are contracted or relaxed. It was previously felt that the hip was in maximum extension in the standing rest position and that man more or less hung from the powerful iliofemoral ligaments. However, this is not the case. From the position adopted by the hip in the standing rest position, the joint can be extended a further 10–15° before maximum extension is reached. The trunk is balanced on the heads of the femur in a more unstable position, and muscular collaboration in the joint's fixation may, therefore, be expected.

In the upright rest position the common centre of gravity for the trunk, head, neck, and upper extremities is at the anterior edge of the lower surface of the eleventh thoracic vertebra. The positional relationship between the vertical line through this point, the trunk's vertical line or weight line, and the hip's transverse movement axis has been reported in several studies to be such that the vertical line passes immediately behind the joint axis. The trunk's weight line should, thus, just about coincide with the body's weight line.

However, according to radiological investigations recently made into the trunk's vertical line, it in fact passes through or immediately ahead of the hip axis. Thus, information on positional relationships between the gravitational line of direction and the hip's movement axis vary somewhat. This may be ascribed to varying methods of study, individual anatomical variations in test subjects or failure to accord sufficient consideration to postural sways. In any case, the trunk's vertical line and the hip's transverse movement axis pass very close to one another or even cross one another. It is not possible to say *a priori* if the joint's postural muscles should be sought among the hip's flexors or extensors.

As previously mentioned, neither the rectus femoris, sartorius, nor the hamstring muscles are actively engaged in the standing rest position. It has also been demonstrated electromyographically that the gluteus maximus is also inactive. This muscle, so powerfully developed in man, is very definitely connected with man's upright posture but it does not collaborate directly in balancing the body in a standing rest position. However, the gluteus medius and tensor fasciae latae are active. The function of these muscles should primarily be to counteract lateral postural sways. But they are also thought to collaborate indirectly in the stabilization of sagittal movements. The iliotibial tract, which lies lateral to the trochanter major, presses the proximal end of the head of the femur, when the tensor fasciae latae is contracted, in a medial direction towards the joint cavity. Since the gluteus medius is attached to the trochanter major and this bony prominence lies in line with the hip's transverse movement axis, the muscle contributes towards fixation of the movement axis.

It is difficult to determine whether or not the gluteus medius and tensor fasciae latae make direct contributions towards sagittal stabilization of the hip by acting antagonistically against gravitational force. The direction of power of these muscles in relation to the movement axis cannot be shown definitely because of *inter alia* the difficulty in topographically indicating the exact position of the movement axis. However, the tensor fasciae latae should be far enough away from the movement axis so that one may assume flexional action by it on the hip and an ability to counteract gravitational force if its line of direction is dorsal to that axis.

The gluteus medius presents a more complicated picture. Mechanically speaking this muscle's anterior part can flex the hip at the same time as its posterior part acts as an extensor on the hip. Thus, this single muscle should be able to stabilize the trunk in its otherwise very unstable position. Because of postural sways, the trunk's vertical line moves forwards and backwards. At one moment it is ahead of, and then an instant later, almost behind the hip axis, intersecting this axis in between. It seems likely that the fluctuation found in the degree of activity in different parts of the muscle may be ascribed to positional changes in gravitational force. If gravity acts to flex, activity increases in the muscle's posterior part and decreases in the anterior part. If gravity acts to extend, activity increases in the anterior part, etc.

If, from a standing position, the trunk is tilted forward slightly, about as much as when one stands at attention, the biceps femoris is activated. The reason for this muscle engagement is because gravitational force is displaced so far ventrally that it exerts a flexional effect on the hip, and the hip's extensors are activated to counteract this moment.

It has been electromyographically demonstrated that the iliopsoas is as a rule active when a person stands in an upright rest position. The iliopsoas would then be a postural muscle.

Judging from hitherto published electromyographic studies of the adductor musculature, it would appear that this musculature does not actively participate in the maintenance of the standing rest position. It is not known whether or not the transverse hip muscles are active in the standing rest position. Their function in different positions and movements has not been studied electromyographically. However, it would not be surprising if a muscle such as the obturatorius externus, for example, with its transverse course parallel to the neck of the femur, were a postural muscle with the task of fixing the condyle in the socket.

This muscular bracing of the hip in the standing rest position appears to occur intermittently in such a manner that certain muscles among both the joint's flexors and extensors may be in action as postural muscles for shorter or longer periods of time, depending on whether or not gravitational force acts to flex or extend the joint.

The trunk's gravitational force need create no moments around the joints of the lower extremities for the sake of lateral stability. If the hip is fixed, passively or muscularly, in the manner described above and the standing rest position is symmetrical, the weight of the trunk is transferred to the two femurs as two equally large, perpendicular forces. The trunk and the thighs then form a mechanical unit and this unit, in turn, is supported on the lower legs. If the femur's mechanical axis in the frontal plane has a vertical course, as reported in the literature, and the extension of this axis coincides with the mechanical axes of the lower leg and the talus, body weight fails to produce any moment in the frontal plane around the knee and ankle. The gravitational force's line of direction coincides here with the forces of reaction.

However, moments develop if the gravitational force does not have the same line of direction as the reaction force. This is the case in the posterior talocalcaneal joint. In Figure 25, we can see how gravitational force $V$ does not have the same line of direction as the equivalent force directed upwards from

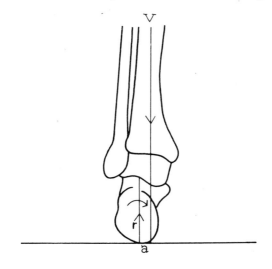

FIGURE 25. Schematic representation of positional relationships between gravitational force, $V$, and the floor's reactive force, $r$, against the heel.

the floor. A moment develops on the calcaneus. This moment, *V.a*, must be compensated for by an equally large (if friction is negligible) moment in the opposite direction so that the position of skeletal parts is retained without the development of pronation. It is impossible to say if it is the soleus muscle or the medial ligaments which prevent pronation, as their forces cannot be measured separately. There must certainly be other passive forces exerting a stabilizing effect. On the other hand, it has not been possible to demonstrate any postural activity in any of the lower leg muscles passing behind the medial malleolus, i.e. the tibialis posterior, flexor digitorum longus, and flexor hallucis longus.

### CONNECTIONS BETWEEN SPINE AND PELVIS

The sacrum, keystone of the spine, is an integrated part of the pelvis. The connection between the sacrum and hip bone is the sacro-iliac joint, a joint in which the applied surfaces are almost interlocking in character. This solid skeletal connection is reinforced by a powerfully developed ligamentous system. The free spine rests on the upper surface of the body of the first sacral vertebra, Figure 26. This load presses the sacrum in like a wedge between both hip bones. The interosseous sacro-iliac ligaments, in particular, which run downwards in a medial direction from the hip bone to the sacrum, are tensed. Pressure force on the sacrum is accordingly transformed into forces which pull both the hip bones in towards the midline and grasp the sacrum like pincers, keeping it from being displaced downwards any further. The line of direction of pressure force passes somewhat anterior to the sacro-iliac joint and acts with a moment on the sacrum. As a result of this moment, the upper part of the sacrum tends to rotate forward and down and its lower part to rotate backwards and upwards. However, this rotational movement is prevented by the powerful sacrotuberous and sacrospinous ligaments.

The fixed connection between the sacrum and hip bone means that if the position of the pelvis is altered, the position of the sacrum is also altered and, accordingly, the shape of the spine.

The invertebral disc between the sacrum and lowest lumbar

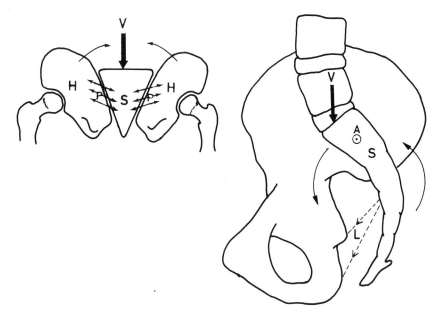

FIGURE 26. Some significant factors in the mechanics of the pelvis.

S: sacrum; H: hip bone; V: spinal pressure on the sacrum's upper surface; L: sacrotuberous and sacrospinous ligaments; P: sacroiliac ligaments; A: movement axis of sacroiliac joint. (*See* further in the text.)

vertebra is much thicker ventrally than dorsally. This wedge shape compensates for the lumbar vertebra's tilt so that the lowest lumbar vertebra is just about horizontal in the standing position. A strong band arises from the tip of the transverse process of the fifth lumbar vertebra, the iliolumbar ligament, with an insertion on the posterior part of the iliac crest. The ligament is in direct contact with muscle fasciae of the quadratus lumborum, psoas major, and psoas minor.

THE SPINE

The twenty-four vertebrae of the spine are arranged one on top of another and joined by intervertebral discs, ligaments, and synovial joints into an S-shaped tube. The task of the spine

71

as a weight-bearing organ may be seen in the fact that the lower the location of a vertebra in the spine, the larger it is. This successive cranio–caudal increase in size is not found in any other vertebrate animal. The bone lamellae in the vertebrae are orientated in a cranio–caudal direction as an expression of the main direction of loading.

The spinal column may be considered to act as a flexible segmental column with both an intrinsic and an extrinsic stability. The intrinsic stability is provided by the alternating rigid vertebrae and elastic discs which are bound together by the systems of ligaments. It has been shown that the critical load for the isolated ligamentous spine is only two to three kilograms. In order to resist the loads and forces to which the spine is exposed, it is obvious that the spine, even during quiet standing, must be kept in position by other and active forces. This extrinsic stability is provided by muscles.

The intervertebral disc, the structure which primarily takes up the load from overlying vertebrae and transfers it to the next underlying vertebra, consists of nucleus pulposus, annulus fibrosus and end-plates, Figure 27. Collagenous fibrillae in the annulus fibrosus run in a gentle spiral from one end-plate and vertebra to the next end-plate and vertebra. The fibres are arranged in concentric lamellae around the nucleus

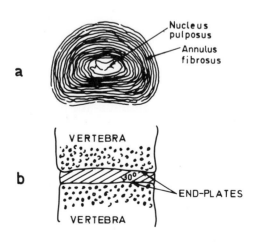

FIGURE 27. Intervertebral disc: (*a*) from above; (*b*) from in front.

pulposus. In each such lamella the fibrillae are approximately parallel to one another and form a 30° angle with the endplates. The angle is open to the right and left alternately in alternate lamellae. Thus, the lamellae cross one another in a series of X's. As the collagenous fibrillae are the elements in the disc which take up tensile force, horizontal tensile forces and tensile forces which form an angle of between 0–30° with the end-plates are those which can be best withstood. The tensile strength of the annulus in a normal disc has been calculated to be 0·9 kg/mm² with horizontal loading but considerably less with vertical loading. Collagenous fibres connect the vertebrae to one another. By their arrangement, they limit horizontal rotational movements and displacements which may arise between vertebrae.

The nucleus pulposus consists of a mucoprotein jelly crisscrossed by a three-dimensional network of collagenous fibrillae. In sections from young individuals it protrudes above the cut surface of the disc. This has been taken as a sign of inner pressure. Discometry now makes it possible to measure pressure in the nucleus pulposus itself. When stress is applied to a normal or slightly degenerated disc, the pressure per unit area is about 50 per cent higher in the nucleus. This is probably due to the elastic resistance of the fibres of the annulus. The nucleus may be considered to be subject to Pascal's law of fluids and consequently is hypothetically incompressible. Thus, the normal disc could be thought of as a kind of rubber tyre with a relatively high internal pressure.

On each surface of the disc above and below there is a thin layer of hyaline cartilage. These layers are called the endplates.

Two force systems act antagonistically within an intervertebral disc. Intra-disc pressure in the nucleus pulposus attempts to tension the fibres in the annulus and separate the vertebrae from one another. This expansive force is counteracted by the annulus and the longitudinal ligaments connecting the vertebrae. Even if the nucleus pulposus is incompressible, it can still be made to change shape and position to some extent. Therefore, the normal disc acts as a shock-absorber and the entire spinal column with discs and ligaments as a flexible rod.

If a motion segment, here meaning two successive vertebrae, is exposed to a vertical load with a direction of application perpendicular to the end-plates and through the intervertebral disc's geometric centre, then a minor change in disc shape can be produced. A normal disc can be compressed by 2·5–3 per cent of its unloaded height. In order to produce such compression (approximately 1 mm in a lumbar disc) without any simultaneous sagittal or frontal disc tilt but with minor bulging of the annulus amounting to about 0·5 mm, an external load of approximately 100 kg or 5 kg/cm² disc area is required. If the load is increased further, the form of the disc remains unchanged but minor but measurable deformation of the vertebra begins to develop. It has been estimated that a normal disc can take a load of up to 500–600 kg before its weakest part, the end-plates, are fractured.

The mechanics of movement in the spine with its ligaments, discs, and synovial joints are extremely complicated and have yet to be explained in detail. For all practical purposes it is possible to divide the kinds of movement into three basic types: ventral flexion and dorsal extension, lateral bending to the right and left, and lateral twisting to the right and left. Movement capability varies greatly in different areas, mainly because of the design and arrangement of the intervertebral synovial joints. These are plane joints supplied with slack joint capsules. In the cervical spine, the joint surfaces lie on a sloping plane through the intersection of the horizontal and frontal planes. In the thoracic area, the joint surfaces are almost frontal and in the lumbar spine almost sagittal.

Intersegmental movements, i.e. movements within every individual motion segment, are relatively small, but the spine's total range of motion becomes considerable if the movements of the different segments are added together. The movements are mainly rotational motions around movement axes common to the disc and intervertebral joints. Since rotational motions are most often combined with minor translation movements, the movement axes are momentary. One may assume that the movement axes pass approximately through or close by the nucleus pulposus. The movement axis which is perpendicular to the joint surfaces of the intervertebral disc may be considered to be the optimum axis for each respective vertebral connection. This means that the optimum movement axis in the lumbar spine is a transverse position with a forward and backward inclination as the essential type of movement in this part of the back. In the thoracic spine, lateral bending around a sagittal axis is the dominant movement and in the cervical spine a combined lateral twisting and lateral bending around an axis running forward and down.

The spine's shock-absorbing ability may also be due to the elastic ligamenta flava which bind together the laminae of adjoining vertebrae. In an upright standing position, these ligaments are tensed and produce considerable elastic resistance to changes in form.

Curves in the spine, lordoses and kyphoses, also contribute to spinal elasticity. But these spinal curves also have a biomechanical effect. A pillar of elastic material which tolerates a certain loading with one curve can bear twice as heavy a load as a pillar with four curvatures in pairs arched in different directions. Pressure and tensile force are distributed among different parts. Elastic equilibrium becomes more stable, and the demand for energy-consuming active musculature declines.

As previously mentioned, the trunk's centre of gravity in an upright standing rest position is on a level with the eleventh thoracic vertebra. The vertical line through this point, the trunk's vertical line, passes in most cases just ahead of the fourth lumbar vertebra's transverse movement axis. This means that gravitational force not only produces pressure forces on the vertebra but even moments around the spine's movement axes, Figure 28. These moments act to bend the spine forwards. The spine's passive forces are not enough to counteract this and muscular activity, as already mentioned, is required. It has also been shown with electromyography that the deep back musculature, the erector spinae, are usually active in the symmetrical standing position while their antagonist, the rectus abdominis, remains passive.

However, in certain rather slightly built youths the lumbar part of the erector spinae has often not been found to demonstrate any activity while the rectus abdominis, on the other hand, is active. In these cases, the trunk's vertical line should pass behind the lumbar vertebrae's transverse movement axes.

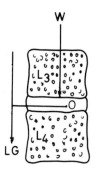

FIGURE 28. The intervertebral disc between the fourth and fifth lumbar vertebrae. Ventral side to the left.

LG: line of gravity; O: transverse movement axis; W: pressure force on the disc.

(*See* further in the text.)

Anthropometric analyses performed on several hundred boys showed that in about every fourth boy the trunk's vertical line passed behind the fourth lumbar vertebra's transverse movement axis. In addition, the analyses showed that between the ages of 15–23 years, the thoracic kyphosis tends to be accentuated while lumbar lordosis becomes less pronounced. If this is the case, the trunk's vertical line should be shifted somewhat in a forward direction during this period.

Electromyographic studies of normal adult material almost always shows that the erector spinae is the postural muscle of the spine in the standing rest position and that it, together with the spine's own passive forces, counter-balances gravity's forward bending moment.

If a person leans forward slightly, then the forward rotational moment is greater, which is also reflected in increased activity in the erector spinae. But if a person bends forward so much that the trunk becomes horizontal, forming approximately a 90° angle to the thighs, activity in the erector spinae terminates, strangely enough, Figure 29. In such a forward bent position, only the back's passive forces collaborate in balancing the trunk, i.e. the back's ligaments, intervertebral discs, joint capsules, and passive elements in the musculature.

But there are also other forces contributing to the balancing of the trunk: intra-thoracic and intra-abdominal pressure.

When a person bends forward, abdominal muscles are activated and often even the intercostal muscles whereby the abdominal wall and wall of the chest become more or less inelastic to pressure. The abdominal cavity and thoracic cavity come to act as closed fluid and air-filled containers respectively. An essential part of the trunk's forward rotational moment will be captured by these 'closed containers'.

In the coronal plane the erector spinae also act to stabilize the spine through collaboration or interplay between the erector spinae of the right and left side. Certain parts of the oblique abdominal muscles have also proved to be in action

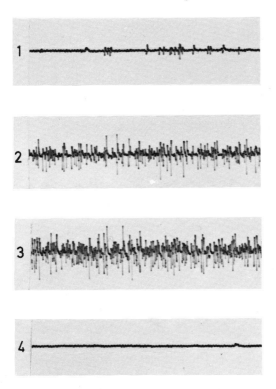

FIGURE 29. EMG from erector spinae.

1: Standing rest position.
2: Standing, approximately 30° forward lean.
3: Standing, approximately 70° forward lean.
4: Standing, approximately 90° forward lean.

in the standing rest position. As far as we know, no electromyographic studies have yet been made of the quadratus lumborum, presumably because of its inaccessibility to electrodes. But in view of its topographical site, it probably contributes actively to lateral balancing of the trunk.

Disorders of the back are common. Even if the real reasons for these disorders are still not known, we do know from experience that prolonged forward-bent body positions and lifting done with a bent back (in particular with the back bent so much that activity stops in the erector spinae) accentuate and often provoke such back disorders. For these reasons, many warnings have been issued about working positions with forward bending, and heavy lifting should be performed with bent knees using only minor forward bending of the trunk and retained lumbar lordosis. The following table may provide an indication of the magnitude of tolerable manual lifts:

| | Isolated, brief lifting | Repeated lifting, 10–30/min | Lifting with 1 min duration |
|---|---|---|---|
| MAN Age 30–40 | approx. 63 kg | 50–36 kg | 32–27 kg |
| WOMAN Age 30–40 | approx. 42 kg | 33–24 kg | 21–18 kg |

Even if the erector spinae in many functional contexts may be regarded as an integrated muscle, one should not forget that it is built up of a number of muscular systems with different anatomical arrangements. Using wire electrodes, it was found that the spinalis and multifidus were always much more active in the standing rest position and in sagittal movements, i.e. bending forward and back, than the longissimus, and the longissimus was, in turn, much more active than the iliocostalis. Since the spinalis and multifidus are attached to the vertebral spines, the iliocostalis to the ribs and the longissimus to the transverse processes, it would appear that muscle systems with the longest levers are those which are primarily engaged; muscles with short levers are engaged secondarily, i.e. when greater power is required. The same principle would also appear to apply in counter-balancing unilateral loads. If a 10 kg weight is held in the right hand, for example, the greatest activity is registered in the iliocostalis on the left side, i.e. in the musculature farthest from the movement axis. According to some authors, the deep back musculature does not collaborate in pure side bending, i.e. bending in the frontal plane, while others feel that it does. But as a rule, a person bends forward somewhat in conjunction with bending to the side, and perhaps it is this forward bending which causes the activity often registered in the deep back musculature with side bending. In twisting to the side, the erector spinae is bilaterally active with activity especially great in the iliocostalis on the ipsilateral side and in the multifidus on the contralateral side. Abdominal musculature also collaborate in twisting.

Someone once said that the best protection for our poor joints is a good musculature. If the back and abdominal musculature, the muscle corset stabilizing the spine's system of joints, is to be strengthened, the same training principles apply as in all strength training: to engage as many muscle fibres as possible in every training contraction. The myograms in Figure 30 indicate the training movements which are most effective, i.e. the movements in which muscle activity is greatest.

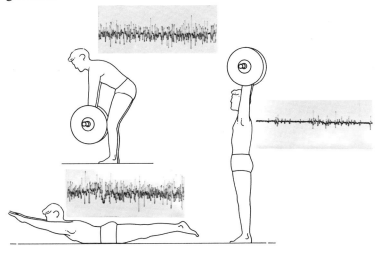

FIGURE 30. EMG from erector spinae showing load on the muscle in different situations. Weight lifted = 45 kg.

In addition to the aforementioned muscles with a postural effect in the standing position, there are still other muscles which are in constant action in this position without really having anything to do with posture. The supraspinatus, the dorsal part of the deltoid and the upper part of the trapezius are such muscles. Without muscular collaboration, the shoulder joint's slack capsule and ligaments would be unable to support the arm and retain the head of the humerus in the shallow joint socket. However, all the arm and hand muscles remain passive when the arm dangles at the side. Finally, it might also be mentioned that the temporalis is the mandible's postural muscle.

Even if all body muscles have not been subjected to electromyographic study and it is not known with certainty if there are any more postural muscles than those mentioned above, one can safely say that only a few of the body's approximately 300 muscles are engaged in the maintenance of the standing symmetrical rest position.

# 4 Sitting

There are many ways of moving from a standing to a sitting position. However it is done, flexing of the hips and knees are the dominant movements. The essential course of events is depicted when a person stands in a symmetrical position in front of a stool, as in Figure 31 and then sits down without using the hands for support. Since the vertical line through the body's centre of gravity must intersect the supporting surface the whole time for the sake of balance, a person must lean forward when bending the knees. The forward lean is produced by flexion in the hips and forward bending of the lumbar spine. The latter movement is displayed in a straightening or kyphosis of the lumbar spine. All these movements conduct the buttocks towards the chair seat. When the buttocks reach the chair the body's supporting area accordingly increases through contact between chair seat and buttocks, and the trunk may once again be raised so that an upright sitting position can be adopted. This may be achieved either by rotating the trunk with a forward flexed spine around the ischial tuberosities or by extending the back, i.e. backward bending in the lumbar spine with the pelvic position maintained, or, which is most common, by combining both these movements.

How large a part of the 90° angle formed by the trunk and thighs in the upright sitting position is due to flexion at the hip joint and how large a part to backward tipping of the pelvis, i.e. by movements in the lumbar spine, depends mainly on the lumbar spine's mobility and varies, therefore, from individual to individual. If changes in the position of the sacrum's upper end or base is used as a measure of pelvic

81

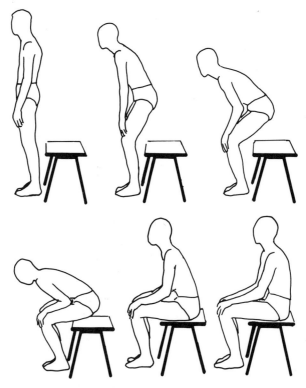

FIGURE 31. Transition from upright standing to upright sitting (according to film sequence).

tipping, then pelvic tipping backwards would amount to about 40° in most healthy adults. Thus, the remaining 50° would be provided by flexion at the hip joint. In the upright standing position, the base of the sacrum leans forward and down and forms a 40° angle, Figure 26, with the horizontal plane. In the upright sitting position, the base of the sacrum is almost horizontal.

The reason for backward tipping of the pelvis is probably because the hips' extensor muscles resist the stretching to which they are exposed when the hip is flexed. They pull the pelvis backwards towards the legs fixed to the chair seat. If the muscles' degree of stretching is reduced by, e.g., bending the knees more and tucking the feet under the chair, causing the

82

muscles' distal attachments to move closer to their proximal attachments, pelvic tipping can be reduced and the lumbar spine permitted to resume lordosis. Under these conditions, as everyone knows, it is easier to sit with a straight back. The forward lean of the trunk, which is included in the pattern of movements in the transition from a standing to a sitting position, leads to a considerable load on the erector spinae, Figure 32. If the hands are supported on the knees, a large part of this load can be transferred to the shoulders.

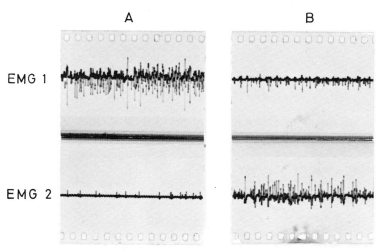

FIGURE 32. EMG 1 from erector spinae, EMG 2 from latissimus dorsi.

A: When sitting without support for the hands.

B: When sitting and supporting the hands on the knees.

When sitting, as when standing, the body's vertical line through the centre of gravity must cross the supporting area. The body's supporting surface is much greater in sitting than in standing. When a person is sitting down, it extends from the anterior part of the foot soles to the posterior margin of the contact area between the buttocks and chair seat. If a person sits with his feet on the floor, the feet bear the weight of the lower leg and, in part, of the thighs. In a severely forward bent position, particularly if a chair is so high that the thighs lean forward and down, a small part of the trunk's weight may be transferred to the feet. However, if a chair is low and the knee angle acute, the trunk's supporting surface, i.e. that part of the chair seat covered by thighs and buttocks, may also carry part of the legs' weight. The trunk's supporting surface, which is the essential sitting surface, is approximately 1500–1600 cm² in area. Thus, the trunk rests on a relatively large supporting surface and the trunk's centre of gravity is relatively low. Accordingly, much less muscular effort should be required to stabilize the body in a sitting position than in a standing position. There need be no load on leg musculature. Balancing the trunk requires a certain amount of muscular engagement, primarily by the deep back muscles and the iliopsoas. The magnitude of this engagement varies, depending on the posture of the back and the degree of pelvic tipping.

DIFFERENT SITTING POSITIONS

When a person sits without support for the back, it is customary to distinguish between a rear, middle, and forward sitting position, Figure 33. In the rear and middle positions, the trunk's centre of gravity is above the ischial tuberosities and the area just to the front of them. In the forward position it is more ventral, somewhere above the middle of the thighs.

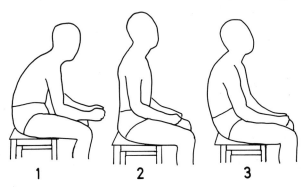

FIGURE 33. Different sitting positions.
1: Forward; 2: middle; and 3: rear sitting position.

In the forward position, the spine forms a total kyphotic arc, just about the position when someone bends forward in a standing position. The hips are flexed considerably while the position of the spine is about the same as in standing. Activity in the erector spinae is insignificant and is often totally absent.

A transition from the forward to the middle position takes place primarily by extension in the hip. This is produced in part by a backward tilting of the pelvis in which the trunk rotates around the ischial tuberosities. At the same time, the spine acquires general lordosis. The middle sitting position is, thus, connected with extension in the back, and this extension becomes more pronounced the less the pelvis is tilted backwards.

If a person sits on a soft surface, the pelvis often remains rotated forward, and the transition from the forward to the middle sitting position then leads to even greater lordosis.

Pelvic tipping backwards is greatest in the rear sitting position. If a person starts from the forward sitting position, the total kyphosis is flattened out when he passes the middle position but is accentuated once again when the rear position is reached. The shape of the spine is generally the same in the forward and rear sitting positions. Only the degree of flexion in the hip distinguishes them. In both positions, load on the erector spinae is insignificant, judging from the degree of activity. The trunk is balanced passively to a large extent. However, the straight back in the middle position demands muscular effort to balance the trunk and fix the pelvis. The iliopsoas is the muscle primarily engaged in this position, but the erector spinae is often active, even if this activity is relatively insignificant.

In order to compensate for backward pelvic tipping with a forward bending of the lumbar spine, a certain amount of mobility is required in this part of the spine. If forward bending is restricted by impaired elasticity in ligaments and muscles or if there are skeletal hindrances, a lordosis or possibly even a straight lumbar spine is retained in the sitting position, even in the rear sitting position. These conditions are the reasons why rest lordosis is encountered more frequently in older rather than younger people. Thus, a direction connection between the degree of pelvic rotation and range of motion in the lumbar spine appears to exist.

There has been a great deal of special interest in the lumbar spine and its change in form in different sitting positions because most back disorders are found in this region of the back. There is much which suggests a close connection between back disorders and unfavourable loading conditions. What shape should the back and the lumbar spine in particular have when a person is sitting, so that load on the back is as modest as possible?

In the upright sitting position with a relatively straight back, loads on the deep back muscles, as already mentioned, are greater than in a relaxed, somewhat hunched position, Figure 34. But reduced load on the erector spinae increases the load on other structures. A hunched position with forward bend-

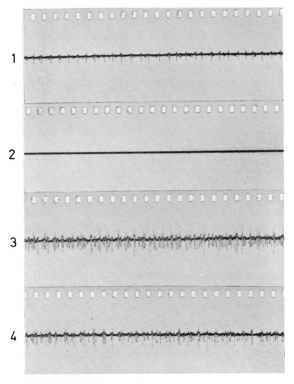

FIGURE 34. EMG from erector spinae.

1: Upright sitting position.
2: Hunched sitting position.
3: Sitting bent forward a little without support for the hands.
4: Same as 3 but with support for the hands.

ing of the lumbar spine imposes increased pressure on the anterior part of the intervertebral discs. The nucleus pulposus is shifted dorsally and load on the dorsal part of the annulus fibrosus increases. Since disc ruptures and disc hernias are usually located in the dorso-lateral part of a disc, presumably because of unfavourable loading conditions, there is reason to assume that the hunched, forward bent position may provoke disc injuries and is, therefore, less suitable than a more upright position, even if the latter requires greater muscular effort.

In any discussion of appropriate and inappropriate back positions, one usually starts with the shape lent to the spine by the intrinsic stability system. Free from external loading, the back's shape is provided by the vertebral configuration and the structure, arrangement, and tensional conditions of ligaments, joint capsules, and intervertebral discs. In the spine's own shape, i.e. the shape generally found in a symmetrically standing man, loading is thought to be optimally distributed amongst the different structures. We do not know for certain if this is, in fact, the case, but there is much to suggest this. Every deviation from this shape causes changes in loading conditions, and we really do not know the physiological limit for such deviations.

However, it is not only distribution of load which varies with different positions of the pelvis and different shapes of the lumbar spine. The magnitude of the load also varies. Recently a measuring device was constructed with which intradiscal pressure can be measured by means of a subminiature pressure transducer. The operating principle is based on the piezo-resistive effect. In a normal disc, it has been found that pressure is about 30 per cent higher in the upright sitting position than in the upright standing position. The lower lumbar discs in the sitting position in adults are subjected to a load of between 100 and 175 kg, while the load in the standing position is between 90 and 120 kg. Tensile forces on the disc have also been calculated. They are greatest in the dorsal part of the disc and may amount to almost 80 kg/cm². These circumstances lend support to the theory that mechanical factors are the reason why disc ruptures generally appear here, even though this is the part of the disc best able to withstand tensile forces.

87

Load on the intervertebral discs is greater in the sitting than in the standing position because, amongst other things, activity in the iliopsoas is greater, and the trunk's moment is greater in respect to the movement axes of the lumbar spine in the sitting position than in the standing position. Backward tipping of the pelvis and straightening of the lumbar spine cause the levers upon which the trunk's weight acts to become longer and the trunk's forward rotational moment greater,

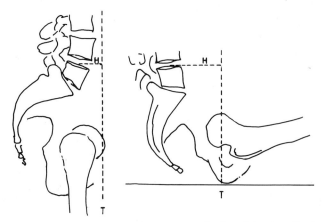

FIGURE 35. The trunk's lever, H, in respect to the fourth lumbar disc in the standing and sitting position.

T = trunk's vertical line.

Figure 35. In order to maintain balance, the backward rotational moment must also be greater. This is shown by increased activity in the erector spinae. Large rotational moments lead to increased load on the intervertebral discs of the lumbar spine. If the backward tipping of the pelvis as well as the flattening of the lumbar spine can be reduced, the trunk's rotational moment becomes less and, accordingly, load on the intervertebral discs reduced. This is why retention of lumbar lordosis in the sitting position is recommended, at least with sitting of long duration without support for the arms.

We have hitherto only discussed symmetrical sitting positions and sitting positions without support for the back so as to clarify the main principles of the mechanics of sitting. But asymmetrical positions are definitely more common than symmetrical ones in our daily life. Even if loading conditions and power interplay are more complicated in the former as the result of torsion and crooked positions, asymmetric sitting positions only differ from symmetrical positions in certain details. Modest twisting and bending to the side usually distinguishes the asymmetrical position from a symmetrical. But even if these deviations from the symmetrical position may seem insignificant, they often cause extremely unfavourable loading conditions and should, therefore, be avoided as much as possible.

### FURNITURE FOR SITTING

A properly designed back rest may facilitate the balancing of the trunk considerably and reduce loads on joints and muscles. Of the multitude of sitting positions adopted in our daily lives, we shall only dwell briefly on three of the most common: (1) sitting at a desk, (2) sitting at a work machine, and (3) sitting in a lounge chair, Figure 36. In all three situations we assume that the chairs have backs and that the height of the seats are well-adapted to the length of the sitter's lower legs. One essential requirement for a good work chair, i.e. for the first two positions, is that a person sitting there should be able to have both feet on the floor with the lower legs vertical (about 90° knee-bending). The height of the chair seat should, thus, correspond to the length of the lower legs. The seat should also be horizontal or lean backwards a few degrees. Its width should be at least 42 cm and its depth 38–40 cm. The region of the trunk's and thighs' supporting surface closest to the back of the knee should not be used. A number of large blood vessels pass here, and their superficial site makes it easy for them to be pinched between the femur and chair seat with blockage of lower leg circulation as a result. Thus, the distal quarter of the thigh's dorsal aspect should not be loaded in sitting.

If a person sits at a desk, he usually sits in a position which is somewhere between the forward and middle sitting positions, but with the difference that there is a back rest and a desk surface upon which to rest the arms to aid in balancing the trunk. Support for the small of the back limits backward

89

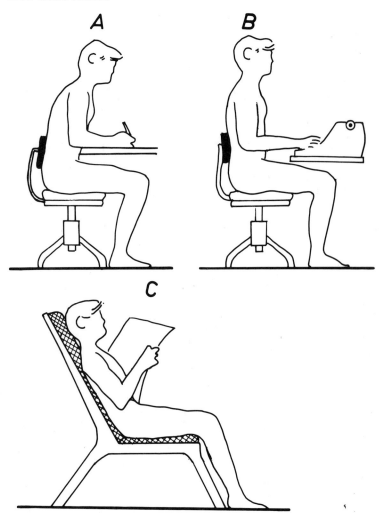

FIGURE 36. Different sitting positions.

A: At a desk; B: at a typewriter; C: in a lounge chair.

tipping of the pelvis and accordingly reduces forward bending in the lumbar spine, i.e. if the back rest is used. But the chair's back rest also has another important function. Interrupting the paper work occasionally, straightening the back and leaning back against the back rest, thereby giving the muscles a rest and temporarily altering load on the back, is a physiological demand which can be satisfied using the back rest.

Resting the arms on a desk top makes it possible for a person to sit for hours leaning forward at a desk. But despite this support, activity in the erector spinae can be stopped only if a person actively presses his arms against the desk top. However, there is a marked reduction in load on the shoulder muscles as soon as the hands are allowed to provide support on the table. It may be useful to point out here that desk height influences the load on shoulder muscles. If a desk is too high, a person often draws up his shoulders and sits with an unnecessarily static shoulder and neck muscle function. If a desk is too low, however, kyphosis of the back develops.

Typists, key-punch operators, typesetters, craftsmen, laboratory assistants, and many other professional groups with manual activities generally sit in a somewhat more upright position but generally without support for the arms. In these jobs the outstretched arms cause great forward torque around the lumbar spine and increase load on the back accordingly. By raising the work field, whether it be keyboard or laboratory bench, and positioning and designing it so that it remains at an appropriate eye-distance even when a person is in an upright sitting position, there is greater opportunity of utilizing the back rest and reducing load on the back. Even if the magnitude of the load on the musculature is greater in this position than in a position with greater forward lean in which there may be support for the arms, distribution of load should be more favourable in the upright position and easily compensate for the increase in loading. Therefore, this position should be preferable as a work position (at least for prolonged work). Various means have been adopted with mixed results in order to reduce load on shoulder and arm musculature in which dull pain and fatigue often develop. Figure 37 shows a support for the left arm fitted to a work chair.

When a person sleeps on his side, he seldom or never has his legs extended. He lies in a more relaxed position with slightly flexed joints. With about a 135° angle in the hips and knees, almost complete passivity is achieved in the muscles, even when a person is awake. Radiological studies have shown that the back generally retains its shape when a person sits on a

FIGURE 37. Work chair with support for left arm.

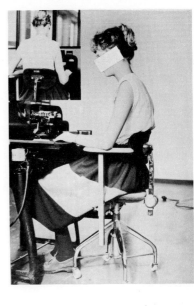

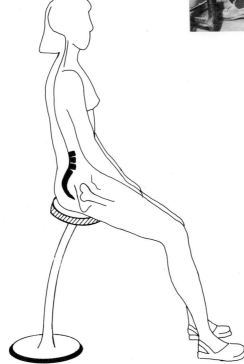

FIGURE 38. A support stool.

support stool from an upright standing position, Figure 38. The hips and knees are then flexed about 135°. This is about the same position adopted when a person wishes to relax in an upright sitting position. He sits on the front edge of the chair seat, leans back and stretches out his legs. If kyphosis in the lumbar region could be prevented by using a support in the small of the back, this position would be the recommended rest position from an anatomical–physiological point of view. Architects and chair manufacturers have also adopted these findings and observations in the design of lounge chairs.

A lounge chair like a work chair should be of proper anatomical and anthropometric design and should also satisfy certain physiological requirements. There is no fixed, ideal sitting position, either when sitting at a desk, typewriter, keypunch machine or when simply sitting and resting. There are always a number of good sitting positions to choose from. The differences between them may appear to be insignificant. But even small shifts in the positions of the joints and the load and tensional conditions of the muscles has a beneficial effect. A good working chair and lounge chair should be designed so that one can easily change body position and adopt different sitting positions.

# 5 The Gait

In any kinematic description of the human gait, the movements of the legs, trunk, and arms dominate the cycle of movement. By exerting a force against the ground in a backward diagonal direction, we accomplish a forward movement of our bodies. As soon as one leg has exerted such a force, i.e. performed a propulsive thrust, it swings forward to a new position in order to perform a new propulsive thrust. Thus, each leg goes through two phases in walking: one phase in which the leg receives support from the ground, i.e. the support (or stance) phase, and one phase in which the leg swings forward, the swing phase. Analagously, one speaks of the supporting leg, the supporting foot, the supporting side, the swinging leg, the swinging foot, and swinging side. Whenever one leg is in the swing phase, the other leg is in the support phase.

But the support phase of one leg does not end the very instant the other leg, after ending its swing, touches the ground, i.e. has its heel-strike. For a few tenths of a second between the swing phases of both legs, both feet are in contact with the ground at the same time. This period in the cycle of movement is called the double support phase, Figures 39 and 40. The duration of this phase varies with walking speed. The faster the walk, the shorter the double support phase. If the gait is so fast that it starts to become running, no double support phase arises. And if we move so quickly that there is a period of double float, i.e. when no foot touches the ground, then we are running.

The support phase begins with the heel-strike. But the other foot takes off only when the entire foot has touched the ground.

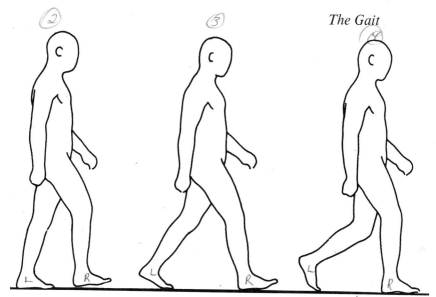

FIGURE 39. Single–double–single support phase in walking.

And when this take-off has terminated, the double support phase ends. With an ordinary gait, the entire support phase lasts about 1 s, the single phase taking about 0·7 s, and the double about 0·3 s.

The gait's cycle of movement is a double step, i.e. both right and left legs have a support phase and a swing phase. At the end of each such cycle of movement, the body parts return to the same mutual positions as at the beginning of the cycle.

The rhythmic movement of the legs also lends rhythmic movement to the trunk, supported by the femoral heads. Figure 41 shows the paths of movement of the head, right shoulder, and right leg projected in a sagittal plane, i.e. in a

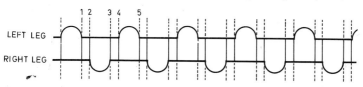

FIGURE 40. Changes between swinging phases (curved line) and support phases (straight line) in walking.

1–2 and 3–4: Double support phases.

2–3: Left leg in single support phase and right leg in swinging phase.

4–5: Right leg in single support phase and left leg in swinging phase.

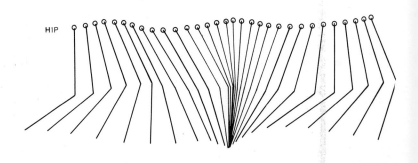

FIGURE 41. Schematic representation of the paths of movement of certain body parts in walking projected in the sagittal plane.

vertical plane parallel to the walking direction. In normal conditions, all people display the same qualitative patterns of movement. Quantitative differences make up the differences between individuals. The various ways people walk are caused by these minor differences.

As may be seen in the Figure, the body parts mentioned describe wave-like movements which for the most part are sinusoidal. In one cycle of movement the body performs two wave movements. There are accordingly two wave crests, maxima, and two wave troughs, minima, per cycle. The wave minima occur during the double support phase and the wave maxima occur during the single support phase. However, these paths of movement are not true sine waves. Thus, the distance between a wave crest and the trough immediately following is less than the distance between this trough and the next crest. Thus, the body's movement forward does not take place at a steady speed.

Wave-like movements are also made on the horizontal

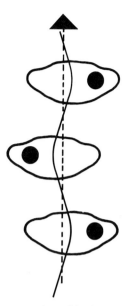

FIGURE 42. Schematic representation of body movements from side to side during a double step. The arrow shows the walking direction. The large black circles indicate the supporting leg.

plane, but these movements have a wave-length which is twice as long as the sagittal path of movement. This is illustrated in Figure 42. Here you see how the trunk moves from right to left and back to the right in a double step. In single support phases, when the supporting leg is vertical, the paths of movement reach their most lateral positions, i.e. the trunk, in relation to the walking direction, is to the far left when the left leg is the supporting leg and to the far right when the right leg is the supporting leg. The lateral turning points in the path of movement occur at the same time as the wave crests in the sagittal path of movement.

If the right and left shoulders in Figure 43 were connected by a straight line and both hips by another straight line, these two lines, here called the shoulder line and hip line respectively, first swing clockwise and then counter-clockwise around vertical axes passing through the middle of the lines. When the shoulder line swings clockwise, the hip line, i.e. the pelvis, swings counter-clockwise and vice versa. When the right leg

97

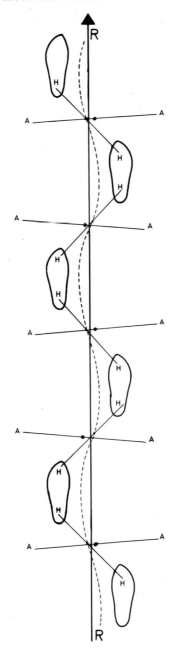

FIGURE 43. Movements of the shoulder line (A–A) and hip line (H–H) in the horizontal plane. The arrow R–R indicates walking direction, the dotted line the centre of gravity of the body. (*See* further in the text.)

swings forward, the right hip also swings forward and the hip line swings counter-clockwise at the same time as the shoulder line swings clockwise. Since the arms follow the swing of the shoulders, the shoulder line may also be taken as an indicator of arm movement. Man moves his arms and legs in walking as four-footed animals move their legs, with a diagonal gait.

The shoulder and hip lines do not merely swing around vertical axes but around horizontal axes, parallel to the walking direction as well. These swings are generally modest in size. Projected in the frontal plane as in Figure 44, the right shoulder and right hip will be seen to be closer together than the left shoulder and left hip when the right leg is the supporting leg and the left leg is the swinging leg. When the body is only supported by one leg during the support phase, the trunk tends to tip towards the swinging side. The pelvis also drops somewhat on this side. But any further drop is prevented by elevation of the shoulder and upper part of the trunk on the supporting side.

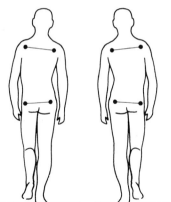

FIGURE 44. Vertical movements of the shoulders and hips in walking.

If the centres of the shoulder and hip lines are connected, a line is obtained which is called the trunk line (Figure 45). If the rotation of the shoulder line and hip line around their centres shows the rotation of the trunk in the horizontal and frontal planes, then the swings of the trunk line illustrate the swinging of the trunk backwards and forwards in the sagittal plane and from right to left in the frontal plane. These swings

mainly take place around a horizontal transverse axis and horizontal sagittal axis respectively through the centre of the hip line. On both the sagittal and frontal planes, the trunk performs three pivots for every double step. The upper diagram in Figure 45 illustrates the pivoting of the trunk line in

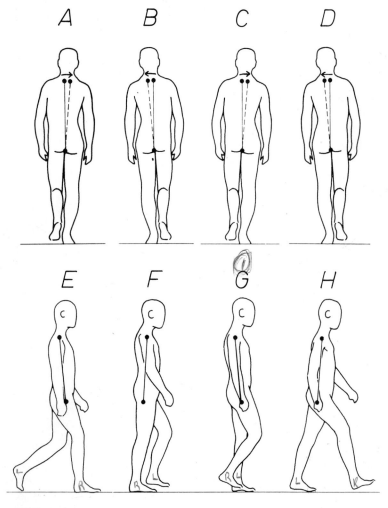

FIGURE 45. The trunk line's swings in the frontal plane (*top*) and in the sagittal plane (*bottom*) in walking.
  (*See* further in the text.)

the frontal plane. In phases A and C, the line displays its maximum right deflection, which occurs every time the right foot is the supporting foot; in phases B and D, the line displays maximum left deflection when the left foot is the supporting foot. The lower part of the Figure shows the swings of the trunk and the trunk line in the sagittal plane. In phase E, the trunk line displays maximum retrograde deflection, which it achieves immediately before the hips and shoulders reach a maximum vertical position and the supporting leg reaches a vertical position. If the right leg is the supporting leg in phase E, the left foot in phase F is just about to touch the ground and the trunk line displays a maximum forward lean. In phase G, when the trunk once again reaches a vertical position, the left leg is the supporting leg and almost vertical. In phase H, the right foot is about to touch down and the trunk line has accordingly swung forward to a new maximum forward lean. These vertical deviations in the sagittal and frontal planes, which the centre of the shoulder line displays in relation to the centre of the hip line, in general only amounts to one or two centimetres.

The movements of the arms, as previously mentioned, follow along with the horizontal swings of the shoulder line. Movements of the arms appear to occur passively and are achieved when the proximal ends of the arms follow the swing of the shoulders, which, in turn, develop as counter-movements in the upper part of the trunk to the torsional movements performed by the pelvis and lower part of the trunk.

PATHS OF MOVEMENT OF THE BODY'S CENTRE OF GRAVITY

Since movements of the body's centre of gravity can be deduced from the direction and magnitude of the forces acting on the body, it is apparent that this point's movements are of special interest. A body's centre of gravity or the centre of gravity of a system of bodies on which internal and external forces act, always moves as if the mass of the entire body were concentrated at this point and all external forces exerted were applied directly to the centre of gravity itself.

In a vertical, standing, symmetrical rest position, the body's centre of gravity is at the forward edge of the second sacral vertebra. But as soon as one begins to walk, and the mutual positions of the body parts change, the centre of gravity also shifts. If the various centres of gravity of body segments are known, one can, at least in theory, point out these centres of gravity, one by one, in picture after picture in a film sequence of a person walking and then calculate for every picture the position of the centre of gravity for the body mass in the position the body happens to be in. The picture series then shows the changes in position and movements in the gait of the person photographed.

As early as the end of the last century, the movements of the centres of gravity were studied in this manner. The results were reported in a co-ordinate system with the position assumed by the body's centre of gravity in a symmetrical, standing position as origo, point *0* in Figures 46, 47, and 48.

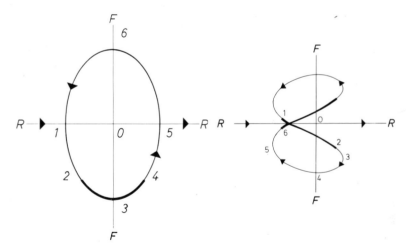

FIGURE 46. (*Left*) Schematic representation of the movements of the body's centre of gravity in the sagittal plane. The arrow R–R indicates walking direction.

(*See* further in the text.)

FIGURE 47. (*Right*) Schematic representation of the path of movement of the centre of gravity in the horizontal plane. Arrow R–R indicates walking direction.

(*See* further in the text.)

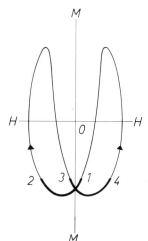

FIGURE 48. Schematic representation of the path of movement of the centre of gravity in the frontal plane, as seen from the rear.
(*See* further in the text.)

The body's movement forward in one step in the walking direction does not occur, as already mentioned, at a steady speed. If the time required for a single step is divided into a series of chronological segments of equal length and the path followed by the centre of gravity during each and every chronological segment is measured, one finds that the speed of the centre of gravity varies considerably during one step. If you imagine a vertical plane, FF, perpendicular to the walking direction RR through point *0*, Figure 46, and that this plane follows the walker and moves forward at a constant speed with a speed equal to the average speed of the step, then the centre of gravity, as a result of its varying speed, will either be ahead of or behind this vertical plane in any one step.

At the same time as the position of the centre of gravity continually changes in relation to this vertical plane, which also continues to move forward just like the centre of gravity, the centre of gravity also swings up and down. The relative path of movement of the centre of gravity, projected in the sagittal plane, acquires the appearance shown in Figure 46. Just before the swinging leg touches the ground, the centre of gravity, in relation to plane FF, is in its most dorsal position, point 1. But the centre of gravity is accelerated forward at the same time as it initially drops and approaches plane FF. During the double support phase, indicated with a broad line

103

between points 2 and 4 in the Figure, the centre of gravity passes the plane at point 3 at the same time as it then begins to move forward and up and reaches, immediately after a propulsive thrust, its most ventral position, point 5. Speed begins to decline, but the centre of gravity still moves upwards. In about the middle of the single support phase, the centre of gravity reaches its highest position, point 6, and then passes the vertical plane, FF, from the front to the rear. The centre of gravity then drops and its speed declines even more and returns to its most dorsal position when the heel of the swinging leg is about to touch down. The movement curve shown in the figure refers to one step. Thus, the centre of gravity describes two such curves in one complete movement cycle.

If both these paths of movement are projected, in the horizontal plane, Figure 47, one will be somewhat to the right and the other somewhat to the left of the sagittal plane through point 0, i.e. the body's median plane, depending on which leg is the supporting leg. In Figure 47, line RR indicates, as in the previous Figure, the body's direction of movement. As with the frontal plane FF, the line passes through the point in which the body's centre of gravity is located, point 0, in a standing, symmetrical position. The more heavily delineated parts of the movement curves indicate double support phases. If the right foot has its heel-strike at point 1, the double support phase lasts until point 2, at which time the left leg concludes its forward thrust. The centre of gravity is then shifted obliquely forward and achieves its most ventral position at point 3, thereafter reaching its most lateral position in its slower phase at point 4 and its most dorsal position at point 5. The forward driving power of the right leg goes into action here and the speed of the centre of gravity increases. At point 6, the left foot has its heel-strike and a mirror image results of the described curve but this time to the left of line RR and with the left foot as the supporting foot.

In the frontal plane, the body's centre of gravity in a double step describes a path of movement like the one in Figure 48. In the frontal plane the centre of gravity moves both up and down and from right to left. HH is a horizontal plane and MM the median plane. When the right leg is the supporting leg, the centre of gravity is to the right of MM and when the left leg is the supporting leg it is to the left of MM. The transition from one side to the other occurs in the double support phase. The heel-strike of the left foot occurs at point 1. The double support phase, during which the centre of gravity achieves its lowest position, then lasts until point 2. The right leg's swing phase and the left leg's single support phase begins here. The centre of gravity is shifted more and more to the left to facilitate body balance, as the body's area of support is only made up of the left foot, and reaches its most lateral position when the entire foot is in contact with the ground. When the supporting leg reaches a vertical position, the centre of gravity reaches, as previously mentioned, its highest position. The centre of gravity has then begun to approach the median plane. When the centre of gravity once again drops during the latter part of the left leg's supporting phase and the right leg's swinging phase, it begins to approach the median plane, MM, which it passes immediately after the heel-strike of right leg, point 3. When the right leg then becomes the supporting leg, the centre of gravity describes a corresponding path of motion to the right of plane MM.

JOINT MOVEMENTS IN WALKING

Flexion and extension of the hip are the most apparent joint movements in the gait. In the double support phase, one hip is extended while the other is flexed. In Figure 39, the left hip is extended and the left leg just about to perform a forward thrust. The right leg is then carried forward and the right hip flexed. Torsion in the lower part of the spine rotates the pelvis around a vertical axis, causing the pelvis and hip line to form a certain angle with the frontal plane in this phase of the gait. The longer the step taken, the greater this torsion. Since the longitudinal axes of the feet are kept parallel or nearly parallel to the direction of travel, this means that a minor medial rotation must be performed in the hip of the rear leg and a minor lateral rotation in the hip of the forward leg, thereby compensating for pelvic rotation.

In the forward thrust, the hip of the rear leg is extended even more, at the same time as the degree of forward leg flexion declines. When the pelvis rotates first forward and outward and then forward and inward during the swing phase, at the same time as the pelvis is moved forward in its entirety in the walking direction, the swinging leg does not follow along completely with this pelvic rotation; its swinging movement proceeds more in parallel with the walking direction. This means that there is adduction, i.e. movement in towards the mid-line of the leg in the beginning of the swing phase and abduction, i.e. movement outwards from the mid-line of the leg at the end at the same time as there is flexion movement in the hip. An initial medial rotation, accompanied by subsequent lateral rotation of the hip, makes it possible to carry the foot forward, despite pelvic rotation, so that its longitudinal axis remains nearly parallel to the direction of travel. During the entire support period there is extension of the supporting leg's hip at the same time as there is medial rotation. The shift of the body's centre of gravity towards the supporting side is achieved partly by an adduction of the supporting leg's hip during the first part of the support phase. In the latter part of the support phase, when the centre of gravity moves medially, there is, on the other hand, abduction in the joint.

Movements in the knee during the support phase are insignificant. At the heel-strike, minor flexion of the joint may occur, and if the propulsive thrust is powerful there may also be extension. But in general, the knee remains relatively fixed during the support phase, Figure 41. Immediately after the propulsive thrust, the knee is flexed and remains flexed during most of the swing phase. However, the magnitude of the flexion angle changes successively. It is smallest during the first half of the swing phase and approaches 180° at the end of the phase. Extension of the knee begins at about the middle of the swing phase when the lower leg's movement forward is more rapid than the thigh's. With a gait with long steps, the knee is just about extended when the heel touches the ground.

Bending of the knee reduces the effective length of the swinging leg and its moment of inertia. In addition, the risk of the foot crashing against the ground is eliminated in forward swinging. The plantar flexion, i.e. sole-down movement which occurs at the moment of propulsive thrust, terminates or may even change into minor dorsal flexion towards the tibia during

106

the first part of the swing phase. During the latter part of the swing phase, when the foot is on the way to touch-down, plantar flexion begins which is rapidly concluded in the first tenth of a second of the support phase. It is within this rapid plantar flexion that the forefoot is put down once the heel establishes contact with the ground.

The support phase is initiated by the heel, and then often by its lateral part coming in contact with the ground, and is concluded by the forefoot and toes lifting from the ground, Figure 49. The movements of the foot during this phase occur mainly in the ankle joint and in the metatarso-phalangeal joints. After plantar flexion at the very moment of heel-strike,

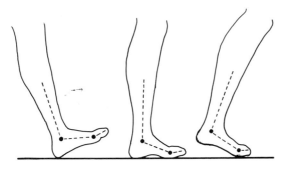

FIGURE 49. Movements of the foot in the support phase.

dorsal flexion in the ankle joint takes over which then continues during the greater part of the single support phase. In the double support phase immediately following, when the foot takes off and propulsive thrust is performed, dorsal flexion changes into plantar flexion. At the same time as there is this movement in the ankle joint, there is also dorsal flexion in the metatarso-phalangeal joints.

Since both the ankle joint and the metatarso-phalangeal joints are simple hinge joints with axes of movement perpendicular to the foot's longitudinal axis, the foot's take-off should be possible only with the joint movements named, assuming that the foot's longitudinal axis and movement of the body's centre of gravity forward completely coincided with walking direction. Even if most people generally walk with the foot's longitudinal direction in the walking direction, the path of movement of the centre of gravity does not completely coincide with the walking direction. As previously mentioned, the centre of gravity in the support phase first moves laterally towards the support side and then medially. This movement to the side is partly in the hip and partly in the joints of the spine as well as in the joints of the posterior tarsus. This adduction is followed by eversion or pronation in the joints of the posterior tarsus when the centre of gravity is shifted laterally as the foot touches down and the body is moved towards the supporting side; in medial shifting of the centre of gravity, the movements are just the opposite.

From the practical, clinical point of view, the foot's eversion, i.e. weight on the inner edge of the foot, and inversion, i.e. weight on the outer edge of the foot, movements are of special interest, as the foot's degree of stability is essentially reflected in these compound movements. When one here speaks of the foot's inversion and eversion, one refers to medial and lateral rotation of the lower leg together with the talus about the longitudinal axis of the foot. Despite a series of studies, it has not been possible to establish how the movements occur in detail in the various joints, i.e. in the talo-calcaneo-navicular, the calcaneo-cuboid and the transverse tarsal joints.

In a study of the movements between the talus and calcaneus by setting pins in both these bones and then measuring in films the angular changes developing between the pins with foot function, it was found that movements during the support phase of the gait were insignificant. In stable feet with a high arch there was $\frac{1}{2}$–1° and in less stable feet there were 5–6° of movement; in individuals with very mobile, flat feet, movement of between 11–12° was measured.

When the part of the foot in front of the transverse tarsal joint, the Chopart joint, is fixed to a surface as in, e.g. the latter part of the support phase (take-off phase), the foot's rear part can be pronated or supinated when it is raised from the ground. These movements then occur in the transverse tarsal joint. This joint is made up partly of the joint between the calcaneus and cuboid bone and partly of the joint between the talus and navicular bone. Detailed studies of movements in the joint showed that in 'pronated' feet the axes of movement of the joints were parallel, permitting free movement around a common axis. This axis is generally parallel to the axis of the talo-calcaneo-navicular joint but somewhat more dorsal than the latter's axis. In the supinated foot, the axes diverge, and movements in the transverse tarsal joints are limited by movements in the talo-navicular joint and the calcaneo-cuboid joint ocurring around an independent axis. Since the transverse tarsal joint shares the talus head with the talo-calcaneo-navicular joint, any supination in the talo-calcaneo-navicular joint increases stability in the transverse tarsal joint. This foot stability is useful when the foot is subjected to forces greater than those in the forward thrust phase. It should be mentioned here that the powerful calf muscles, the gastrocnemius and soleus, the most powerful take-off muscles, have a supinative effect when the forefoot is fixed to the ground.

KINETICS OF THE GAIT

Scientists have used the chronocyclographic method for both the registration of the body's centre of gravity and the paths and speed of movement of the various body segments as well as for a study of the kinetics of the gait. However, the method is very time-consuming and uncertain for use in calculating the gait's interplay of forces. Values received for the magnitude and direction of these forces are too approximate. The greatest source of uncertainty in the method is the difficulty in indicating and marking the correct positions of the centres of gravity of the body segments.

Newton's Law governing effect and counter-effect, action and reaction, has provided the basis for measuring instruments brought into use in recent years for direct measurement of the forces in effect between the foot and ground. These forces in particular have attracted the greatest research interest, and familiarity with them is of fundamental importance in understanding the kinetics of the gait.

The most reliable measurements are without doubt those made with the aforementioned force plates or step force meter, Chapter 2. These instruments are either recessed in the floor or make up part of the walkway as in Figure 11. The force components usually registered with the force plates in studies

*How Man Moves*

of the gait are vertical forces, horizontal forces in the walking direction, and horizontal forces perpendicular to the walking direction.

The magnitude of the forces exerted by the foot against the ground or the force plate is primarily dependent upon body weight, step speed and the speed of direction of movement of the body's centre of gravity. Figure 50 shows a registration from a walking experiment.

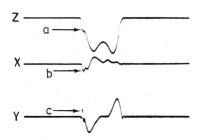

FIGURE 50. Registration of the foot's force against the ground (the force-plate) in walking.
(*See* further in the text.)

The upper trace in the Figure, trace Z, shows the vertical force component during the support phase, i.e. the body's vertical acceleration movements. It has two crests with an intermediate trough. The trace has here been turned upside down to save space. The first crest occurs immediately before the body's centre of gravity is raised to its highest point above the supporting foot, and the other crest occurs immediately after this point. But a little twitch is often, but not always, noted during the climb of the first crest, as shown by an arrow in the Figure. This twitch develops because the leg more or less gives way to the force imposed on it at the heel-strike. The body's acceleration declines for a brief moment, presumably because of bending movements in the joints, primarily in the knee.

In relation to body weight, the vertical force in the trough is always less than body weight, while crests generally display values greater than body weight. However, there are cases in which only the second crest exceeds body weight. During the first crest in these cases, the body is accelerating downwards. If the knee is bent when the foot touches down and the leg is not straightened until the propulsive thrust, the first crest will be lower than the other, occasionally so low that it fails to

110

achieve a value corresponding to body weight. However, if the body is dropped in propulsive thrust, the second crest never comes up to the level of body weight. In certain cases, traces are obtained in which neither of the crests are on a level with the body weight. This occurs if a subject passes over the step force meter with rapid strides and bent knees.

The trough between the trace's two crests develops because the body accelerates upwards. This is due to the fact that the reaction forces, musculature, and ground counteract the body's movement downwards during the first part of the support phase.

The horizontal force component acting in the walking direction is the strongest component of the horizontal forces, trace Y, Figure 50. When the outstretched swinging foot touches down, the body's centre of gravity is behind the touch-down point. The result of this is that the ground's reactive force has a horizontal retrograde component which exerts a braking effect on the body's movement. This effect terminates when the body's centre of gravity passes over the supporting foot. This occurs when the supporting leg is just about vertical. The supporting leg's pressure against the ground then begins to be directed backwards more and more. The leg pushes off. The reactive force acts in the opposite direction, driving the body forward. When the supporting leg is vertical, the foot's departure from the ground begins.

The horizontal, sagittally exerted force curve contains two main constituents: (1) the braking constituent, when the supporting leg exerts a forward force, and (2) the propulsive constituent, when the supporting leg exerts a retrograde force. In many cases, one finds that the supporting phase is introduced by a brief (0·05–0·1 s) retrograde force, arrow c in Figure 50. This force is a result of the heel's thrust force against the ground. It does not occur if the entire foot is set down all at once. If the heel touches down first, the whole foot then rotates around the support point at the same time as the lower leg is moved forward. This movement presumes a forward frictional force on the heel, i.e. an equally large retrograde force on the step force meter. This force terminates when the entire foot is on the ground, Figure 51.

In our account of gait kinematics, we have pointed out that

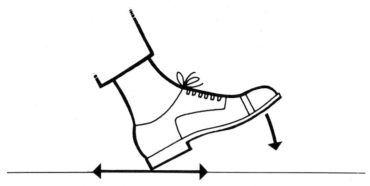

FIGURE 51. The foot at the moment of touch-down.
(*See* further in the text.)

the body's centre of gravity is shifted alternately to the right and left when the right or left leg is the supporting leg. The purpose of such a shift of centre of gravity is to give the body the most stable position possible during the single support phase above the body's supporting surface, i.e. the supporting foot. If the centre of gravity, despite these lateral shifts, fails to lie directly above the supporting foot but somewhat medial to it, a transverse force component develops. As a rule, this component is insignificant, usually only a few per cent of body weight, but values of up to 25 per cent of body weight may occur. Most people have a lateral component during most of the support phase, curve X in Figure 50. However, the supporting phase begins with a brief, medial component.

During the swing phase, when the pelvis rotates around the supporting leg, there is a movement constituent, with reference to a vertical axis, through the supporting leg. When the swinging foot touches down, this movement is retarded by a thrust. It is this thrust which causes the medial crest, b, on curve X at the beginning of the support phase.

At the same time, one might expect to find an equally large medial force on the other foot's X curve, i.e. at the end of the support phase. However, this is not the case because of the following: at the instant of touch-down, the body's movement to the side is laterally directed towards the side of the supporting foot to be. This provides a thrust which acts medially on the coming supporting foot's X curve. This causes a certain reduction in the medially directed force generated by the retardation of the aforementioned movement. Despite this reaction, the propulsive foot's medial force is less than the touch-down foot's medial force. This may be because the axis around which the impulse constituent moves is closer to the propulsive foot than to the touch-down foot. In addition and possibly most important of all, there is a lateral component in the retrograde force of the propulsive foot which helps to move the body over towards the side of the supporting foot to be and which reduces the impulse constituent's medial force component so that this becomes less than the medial component of the touch-down foot.

When the foot's force against the ground is measured with a step meter or, which is really the same thing, the ground's thrust against the foot, it is impossible with this instrument to determine the part of the foot exerting these forces. By using various triggering methods or by photographing the foot from below when a subject passes over a glass walkway, the successive shifts in load from the heel to forefoot and toes during the support phase can be studied roughly.

A more exact and detailed method is to place small pressure-sensitive piezo-electric plates on the underside of the foot. Figure 52 shows a resulting measurement and how the load is distributed on various parts of the sole of the foot during the support phase. First the heel's lateral part is loaded, then its medial part and then, one by one, the fifth, third and first metatarsal bone heads and finally the big toe.

Another way of studying in detail the shifting of the vertical force component during the support phase is to register this

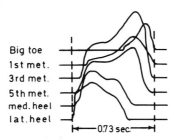

FIGURE 52. Distribution of load at six parts of the heel and foot sole during the support phase (according to Schwartz and Heath).

113

force with the aid of built-in force plates in shoes, Figure 12. Figure 53 shows a registration of the shift of the vertical force and variations in its magnitude during the first part of the support phase while the load is mainly on the heel. When the heel touches the ground the vertical force component first moves dorsally, turning almost immediately to shift ventrally towards the forefoot. This retrograde shift and the vertical power component occur at the same time as the aforementioned horizontal component which occurs in the initial part of the support phase, arrow c on curve Y in Figure 50.

As previously mentioned, the movements of the arms in walking appear to occur passively. It has been found in studies of energy expenditure in walking on level ground that movements in the spine and the shoulder reduce and brake movements transmitted from the legs and pelvis to the upper part of the trunk. In this way, the body's movement forward is

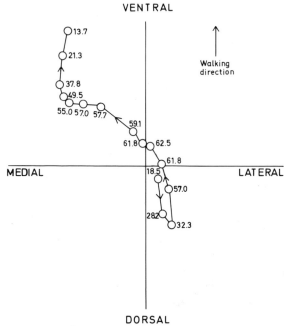

FIGURE 53. The magnitude in percentage of the body weight and shifts in position of vertical heel force during the support phase (according to Wetzenstein).

made more efficiently. Accordingly, the movements of the arms restrict the trunk's rotational movements and smooth out the transition between these movements.

If the back is immobilized and rotational movements of the pelvis and shoulders are eliminated, general energy expenditure increases. With such an immobilization there is a subjective feeling of increased difficulty in maintaining the balance, increased side movement of the shoulders and increased rotational movements in the trunk as a whole.

MUSCLE CO-ORDINATION

The muscles which are actively engaged in the human gait have not yet been fully established. From disturbances in and deviations from the normal gait pattern as a result of partial muscle paralysis, it is possible to assume that certain muscles and groups of muscles play an important role in walking. But it was not possible to follow the muscle interplay in detail until the advent of electromyography. There are still muscles whose function is not known with certainty. These are mainly deep muscles, difficult to reach for electromyographic registration, muscles such as those of the hip and foot.

Despite these gaps in our knowledge, it has still been possible to acquire a relatively good picture of muscle co-ordination in an ordinary gait. The way in which one determines in which phase of the gait cycle a muscle is active may be seen in Figure 54. Figure 55 shows schematically the different periods of muscle activity during a step. As may be seen from the diagram, there is often good agreement between a muscle's mechanical pre-requisites and its true function. But this is not a general rule. Certain muscles remain passive, even if they could participate mechanically. There are also muscles which are active in movements they are unable to produce.

During most of the swing phase, there is extension in the knee and flexion in the hip. It may then seem odd that the hamstring muscles are activated in the latter part of the swing phase. These muscles act antagonistically on both these movements and therefore retard the forward swing of the leg. By retarding the bending of the hip, a large enough part of the thigh's momentum could be transmitted to the lower leg so as

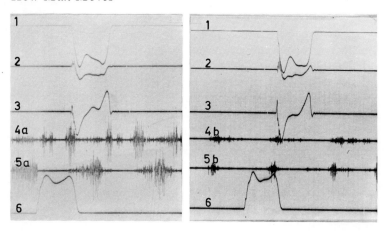

FIGURE 54. Registration of two steps, one after another. Walking direction from left to right.

Curves 1, 2, and 3: left foot's vertical, horizontal transverse, and horizontal sagittal force; curve 6: right foot's vertical force.

EMG 4a: Left tibialis anterior; EMG 5a: left soleus; EMG 4b: left vastus lateralis; EMG 5b: left biceps femoris.

to enable the latter to swing forward to maximum extension of the knee without the aid of extensor muscles. Such passive extension of the knee might conceivably be advantageous from the point of view of energy, but balancing the magnitude of the movement would be more difficult and carry with it a risk of over-extension. By simultaneously activating the hamstring muscles and vastus muscles, i.e. co-contraction of the knee muscles, movement takes place in the joint with satisfactory control, and the joint's stability is improved considerably, which is essential when the foot touches down and a severe load is suddenly imposed on the joint.

As a rule, the knee is bent somewhat when the foot touches down. At the instant of touch-down, the joint is bent still further, dampening the impact against the ground, and ground thrust is transmitted to the body more gently and resiliently than if the joint were extended to a maximum. The fact that this retardation of movements in the hip and knee is not performed by, e.g. the gluteus maximus or the short head of the biceps femoris may be because *inter alia* less energy is required if a piece of work is performed by a two-joint muscle rather

116

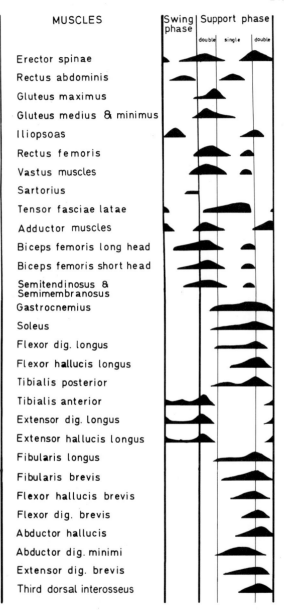

FIGURE 55. Muscle co-ordination in ordinary gait: schematic representation following electromyographic registrations.

than by two separate one-joint muscles.

When the foot touches down, the hamstring muscles act with a closed muscle chain. Their mechanical function is altered accordingly. From acting as flexors of the knee in the swing phase, they become knee extenders in the support phase. From being antagonists of the vastus muscles, they become their synergists.

The powerful adductor musculature, which will be treated as a unit here, displays powerful activity crests during a double step. These occur at the same time as the double support phases. The essential task of this muscle group in walking appears to be to stabilize the pelvis at the supporting leg. Mechanically speaking, this muscle group is able to flex in the hip when the joint is extended and extend in the hip when the joint is flexed. Even if this is not the most important task of the adductor muscles in walking, they still work synergistically with the iliopsoas in the beginning of the swing phase and synergistically with the hamstring muscles at the end of the swing phase when the leg's forward swing is actively retarded. In the frontal plane, the adductors work antagonistically against the gluteus medius and tensor fasciae latae. As both abductors and adductors are active in the first part of the support phase, the adductors contribute towards fixing the pelvis and moderating the interplay of forces between abductors and body weight. The adductors are also mechanically able to rotate the leg on the horizontal plane against the pelvis or, if the leg is fixed, the pelvis against the leg. When the hip is flexed, the adductors rotate the leg inwards. If the hip is extended, it rotates the leg outwards. The most important pelvic rotation on the horizontal plane takes place at the same time as the adductors are active. But since the direction of muscle pull is antagonistic to pelvic rotation occurring, this means that the adductors counteract rotation and are included in the force which counteract the body's torsional movements in walking. Thus, in the instant of forward thrust, the active adductor muscles act repressively towards stretching movements in the hip and towards pelvic rotation forward and medially.

The plantar flexor muscles, the soleus, tibialis posterius and others, are activated during the support phase at the same time as they are extended. This has a favourable effect from the point of view of power. In an extended active state, the muscles can subsequently apply plantar flexion with great force to the foot when the latter takes off, perhaps the most important phase in the entire gait cycle.

In the double support phase body weight is shifted towards the coming support side, and the foot on this side becomes strongly pressed onto the ground. This shift in body weight leads to an eversion movement in the joints of the posterior tarsus and to a tendency to bending in the knee. However, the eversion movement, as well as the bending, is inhibited both by the ligaments and active muscles. The tendency to eversion is counteracted by the supinating and active muscles, the soleus and tibialis posterior, and bending by, in addition to the medial knee ligaments, the active muscles passing medial to the knee, the semitendinosus and gracilis. In addition, the body's centre of gravity remains medial to the supporting leg's various joints, which is why the body mass acts to turn inwards on these joints.

The tensor fasciae latae is occasionally and the sartorius infrequently active in the leg's swing phase. In this phase, there is hip flexion but hip flexion associated with minor medial rotation in the first part of the swing phase and with minor lateral rotation in its latter part. Also the tensor fasciae latae may be active in the beginning of the swing phase; at the end of the swing phase the sartorius may be active. The magnitude of these rotational forces varies from case to case, depending on the kind of gait and the occurrence of activity and degree of activity thereafter. It is these rotational steps and not flexion in the hip to which the activity correlates. Muscles such as the abductor hallucis, flexor hallucis brevis, flexor digitorum and abductor digiti minimi lie for the most part along the foot's longitudinal axis and can produce powerful bending in the transverse tarsal joint, thus playing a decisive role in the stabilization of this joint during the take-off while simultaneously supporting the foot arch.

Even if the ventral muscles of the lower leg are active at the beginning of the support phase, contributing thus to dorsal flexion in the ankle joint, this movement should essentially be a result of the kinetic energy imparted to the body by the other leg in its forward thrust. The task of the muscles should essentially be to stabilize the ankles and control touch-down of the foot so that the forward foot does not fall passively to the ground after the heel has touched down in the initial stage of the support phase. The kinetic energy driving the lower leg forward is moderated by the active soleus and tibialis posterior and by the extension of the knees, which then prevent excessive knee-bending. The joints are fixed actively by muscles and passively by ligaments so that the supporting leg provides a sufficiently solid and stable but, at the same time, somewhat resilient and elastic supporting pillar. This is the reason why muscles are engaged both to facilitate movement and to stabilize the joints. The function of the gluteus maximus, gluteus minimus and, to a certain extent, even the tensor fasciae latae during the support phase is to prevent the pelvis from sinking on the swinging leg side at the same time as they control rotation of the hip inwards and outwards and also to extend the joint.

The result of the bilateral activation of the erector spinae in the final stage of forward thrust is not merely extension of the back but perhaps even more an accentuation of the hip's extension. In this latter movement, the deep back muscles cooperate with the hamstring muscles. This extension of the hip immediately before the swing phase leads to stretching of the iliopsoas which can give the leg a powerful swing forward. The erector spinae also initiates torsion of the spine following forward thrust. This movement, like the leg's forward swing, is ballistic with brief activation of the muscle, the movement thereafter being completed by the momentum of the various body segments.

# 6 Kinesiology of the Arm

While the lower extremities are the body's support and organs of locomotion, the upper extremities are the body's gripping organs, with the ability to grip located in the hand. One speaks of manual labour, hand grenades, hand-ball, hand-stand, handy, etc., the point merely being that it is the hand which grips or touches the object. But these hand functions are to a large extent dependent upon the function of the entire arm, i.e. in almost every gripping movement, as in almost every arm movement, there are polyarticular movements in which all or most of the joints of the upper extremities are involved. The gripping surface of the hand and gripping power are facilitated by the advantageous position of the hand at the periphery of the lever system formed by the upper and lower arm.

The arrangement of the shoulder joint as a ball-and-socket joint with a displaceable socket gives the arm a very wide range of movements. This range of movements is mainly ahead of and lateral to the body and within the eye's field of view. Since the arm has a hinge joint at the elbow and an ellipsoidal joint at the wrist, the hand can reach almost any point within this range of movement. Thus, the hand can reach almost any point on the surface of its side of the body and a large part of the surface of the other body half.

It would be meaningless to try to classify the arm's movements according to any kinesiological norms. The forms of movement, as with the movement functions, are too varied and too difficult to define to permit accommodation in sufficiently well-defined classes, types, or functions.

The movements of the arm are usually anatomically divided into sub-movements, and every joint and its typical move-

121

E

ments regarded individually. By adding up the degree of freedom in various joints or their typical or basic movements, it is at least possible to express geometrically the movement alternatives in the arm's joint system. But this provides no information on the arm's movements in relation to the body.

In business and industry, arm movements are also divided into sub-movements or elements. 'Grasp', 'Transport loaded', 'Hold', 'Release load', etc. are such elements. From the point of view of muscle function it is often useful to differentiate between open and closed arm movements. Open movements are movements in which the arm moves freely, with or without a tool or instrument in the hand, and movements are so to speak referred to the body, the trunk providing support for arm movements. Closed movements are movements made with the hand fixed to an object which will not move, e.g. pressed against the floor in push-ups, or an object whose direction of movement is mechanically controlled, e.g. a car steering wheel or gear lever. However, in open movements the trunk does not have to be fixed but may often follow along with arm movements and contribute to the arm movement's greater efficiency. This is the case, for example, in many throwing movements. In manual precision work, 'movement support' is often transferred distally from the trunk in order to increase precision of movement. Coarse co-movements in the shoulder and elbow are, thus, eliminated. Subtle finger movements are accordingly not just miniature versions of total arm movements but completely different movements restricted to the arm's distal joint system. With closed arm movements, the arm is mechanically connected to an external object so that the object may be designated as the fixed support for the movements. If you sit in a car with your hands on the steering wheel, you are then mechanically connected to the car's steering mechanism, and both links together make up a mechanical system.

If in a standing, upright position, the arms are allowed to hang relaxed at the sides, it is mainly arm weight and the passive elements in muscles, tendons, and ligaments which provide the hand and arm with the characteristic position, the rest position. Only insignificant rest activity can be registered in arm and shoulder musculature with this arm position. The joint surface on the head of the humerus is pointed medially and dorsally and partially upwards, and the glenoid socket of the scapula is pointed laterally and ventrally and partially downwards when the arm is in the rest position. Frequently only the lower half or a small part of the head is in contact with the socket, Figure 56. A perpendicular line through the centre of the contact surface forms an approximately 90° angle

FIGURE 56. Position relationship of the shoulder blade to the upper arm in the limply hanging arm (*left*) and with the arm abducted approximately 130° (*right*).

with the shaft of the humerus. In the horizontal plane, this perpendicular line, together with the frontal plane, forms a forward-outward open angle of 30°, which means that this perpendicular line lies in line with the scapula. The upper end of the humerus, i.e. the head and neck of the head (collum anatomicum) is somewhat retroverted in relation to the lower end of the humerus, Figure 57. This means that the axis of movement of the humero-ulnar joint, i.e. the elbow's flexion and extension axis, forms an angle with the frontal plane

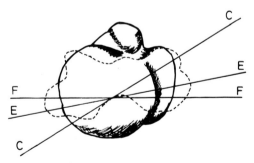

FIGURE 57. The right humerus from above.

Dotted line indicates the lower end of the humerus.

Line F–F: the frontal plane; line E–E: the elbow's flexion-extension axis; line C–C: position of the neck of the humerus.

123

which is somewhat less than 30°. How much less varies from individual to individual. Since this axis of movement more or less passes through both epicondyles, a simple palpation will disclose the relative positions of the axis of movement and the frontal plane. However, the radial (lateral) epicondyle is always ventral to the ulnar epicondyle when the arm hangs relaxed.

The elbow is somewhat flexed in the arm's rest position and the hand is half-pronated with the palmar side facing the thigh. The fingers are gently bent and the thumb's palmar surface faces in the direction of the ulna.

Many arm movements can start from this position of rest. For didactic purposes, it might be useful in an account of the arm's kinesiology to first follow the prevalent pattern and discuss each joint individually. This analysis will be facilitated if you have a rather clear picture of the kinematics and kinetics of the various joints and an understanding of total arm movements such as those encountered in daily life and to which there is reason to return.

## THE SHOULDER JOINT

Movements in the shoulder joint are intimately connected with the shoulder girdle and, in particular, with the motion of the scapula. The scapula which, via the clavicle and both the clavicular joints is connected to the thorax, is initially a stabilizer, so to speak, and the shoulder initially provides movement. In certain pathological conditions, the primary motion-giving function of the shoulder joint is cancelled out and there is reversal of function so that the shoulder joint is stabilized and the arm's movements are made by displacing the scapula towards the thorax. This is a condition often encountered in 'frozen shoulder'. Fixation of the scapula and its displacement along the chest is mainly accomplished by three muscle loops, Figure 58. There are two muscles in each such loop which run from the scapula in diametrically opposed directions: the levator scapulae—lower part of the trapezius, rhomboids—serratus anterior, and upper part of the trapezius—pectoralis minor. Also the sternomastoid and subclavius muscles connecting the clavicle to the trunk are impor-

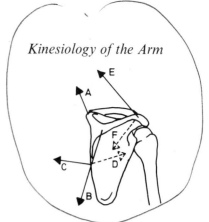

FIGURE 58. The three muscle loops of the shoulder blade.
A & B: Levator scapulae and lower part of the trapezius.
C & D: Rhomboids and serratus anterior.
E & F: Upper part of the trapezius and pectoralis minor.

tant to the mobility of the shoulder girdle but do not take part in the movement of the arm in the same intimate and direct manner as the three muscle loops.

Electromyographic studies have shown that the upper and middle parts of the trapezius, levator scapulae, and the upper part of the serratus anterior are the muscles which actively (even if activity is rather insignificant) keep the scapula stabilized in a rest position. With increased activity, these muscles draw the scapula upwards. Even with rotation of the scapula, these three groups of muscles form a unit. The other unit in the power pair rotating the scapula is formed by the lower part of the trapezius and the lower and larger part of the serratus anterior.

With every movement in the shoulder joint, some part of the three trapezius sections are engaged, even if the distribution of activity among the different sections varies with the arm's direction of movement. Thus, with forward swing of the arm, the entire trapezius is active. However, the upper part is much more intensely activated than the other two parts. In backswing, the activity is evenly distributed over the entire muscle while the middle part is the most active part in abductive movement. Even when the arm is rotated laterally, the middle part is the muscle primarily activated; the lower part is activated somewhat less strongly and the upper part least of all. With medial rotation, activity is apparent in the lower part while the other two parts are only slightly active.

The rhomboids acts like the middle part of the trapezius and is most active in the arm's abduction and least active in the initial stage of flexion.

Two types of fundamental movements can be discerned in

125

the shoulder joint: elevation and rotation of the upper arm. Both these types of movement are almost always combined in the movements executed by the arm in our daily work. Elevation can take place in different directions: forwards, backwards, obliquely forward, forward and up, outwards, etc. The direction of rotation combined with these elevation movements is determined by the arm's starting position and the direction of elevation. It may be of practical use to remember that the head of the humerus, which lies hidden under the musculature, is pointed in approximately the same direction as the easily palpated medial epicondyle.

Even in apparently simple movements, forward swinging and abduction are combined with a rotation of the arm. Included in the abduction of the arm, there is an outward rotation in which the head of the humerus glides forwards and downwards in the glenoid socket so that retrotorsion adopted by the head of the humerus when the arm hangs freely turns into antitorsion when the arm is abducted to a maximum. Minor inward rotation of the arm is generally part of a habitual forward swing at the same time as the head of the humerus glides backwards somewhat.

It is no exaggeration to claim that the interplay within the shoulder musculature in arm movements is very complicated. There are several reasons for this complicated pattern of co-ordination.

The shoulder joint, in which the most important movements take place, is a ball-and-socket joint with a wide range of movement. It has a shallow joint socket and, in relation to the socket, a large spherical ball. The joint surfaces are not congruent, as the radius of the curve of the socket is generally larger than the condyle's. The joint capsule is slack and the ligaments are relatively insignificant. Therefore, participation of muscles is required for fixation of the condyle in the joint socket.

Another condition complicating co-ordination of the shoulder musculature is that the position of the socket changes successively during arm movement. A major part of the arm's range of movement consists of a displacement of the shoulder girdle by movements in the sternoclavicular joint and acromioclavicular joint. With a 180° abduction or a 180° forward and upward swing of the arm, approximately 120° of the range of motion is carried out in the shoulder joint itself while the remaining 60° is achieved by displacement of the scapula along the chest. When the arm hangs limply at the side, the humerus and the spine of the scapula form approximately a right angle with one another. But when the arm has been abducted to approximately 140°, they lie in line with one another. This means that the longitudinal axis of the humerus and its mechanical axis will coincide after initially having been at right angles to one another, Figure 56.

The site and direction of muscle pull in relation to the shoulder joint also changes successively during a movement. Both the dorsal and ventral parts of the deltoid muscle have an adductive pull direction on a limply hanging arm, Figure 59.

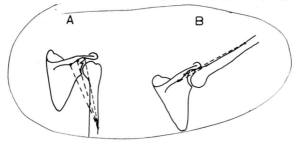

FIGURE 59. The shoulder joint. The dotted lines indicate the dorsal part of the deltoid muscle.

A: A limply hanging arm; B: arm abducted approximately 130°.

But when the arm is abducted, an increasing number of muscle fibres of these muscular parts will end up above the sagittal axis of movement, according to the progress of the movement, and will thus acquire an abductive direction of pull similar to that of the part of the muscle arising from the acromion. Different parts of the same muscle, indeed even parts of the same muscle section, may accordingly be mechanical antagonists to each other. Mechanical conditions are also changed for the muscles with insertions on or near the greater tuberosity as the arm is abducted.

Another example: lesser tuberosity and the medial tip of the bicipital groove are lateral to the shoulder joint's vertical axis of rotation and in front of its transverse axis of movement when the arm hangs limply, but respectively medial and behind these axes when the arm is stretched upwards. This means that the ability of muscles attached to these bony prominences to influence the shoulder's movement is successively changed during the course of movement.

If co-ordination were 100 per cent mechanically functional so that only muscle parts which could achieve motion were activated, this would mean that certain muscles would be gradually brought into or removed from action as their mechanical conditions changed during the course of a movement. As may be seen in the diagrams in Figures 60 and 61, this is not the case. Almost all the shoulder muscles are in action throughout most of an abduction movement and forward upswing, irrespective of whether or not the direction of pull is such that the muscle counteracts or contributes to arm

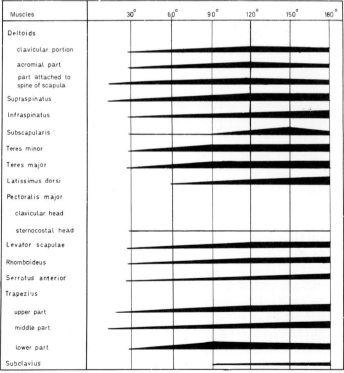

FIGURE 60. Schematic representation of muscle activity in arm abduction.

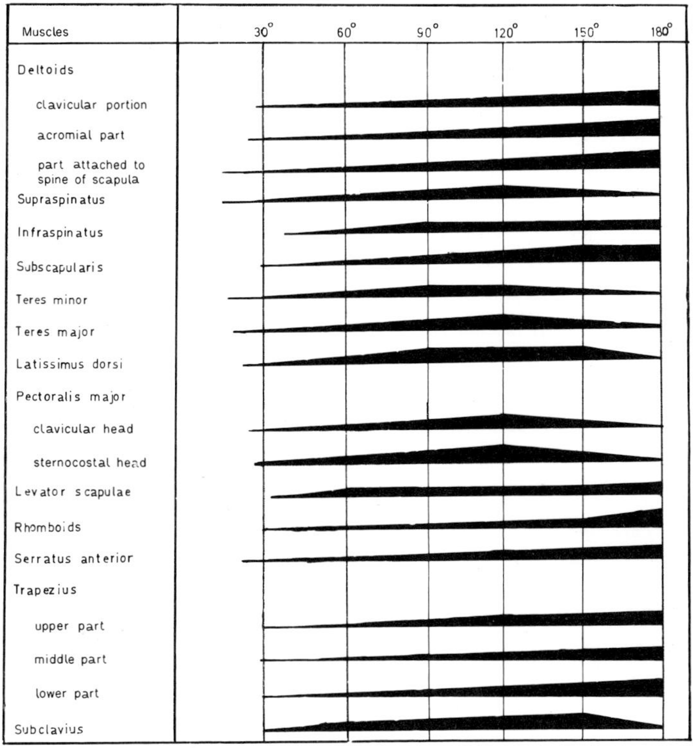

| Muscles | 30° | 60° | 90° | 120° | 150° | 180° |
|---|---|---|---|---|---|---|
| Deltoids | | | | | | |
| clavicular portion | | | | | | |
| acromial part | | | | | | |
| part attached to spine of scapula | | | | | | |
| Supraspinatus | | | | | | |
| Infraspinatus | | | | | | |
| Subscapularis | | | | | | |
| Teres minor | | | | | | |
| Teres major | | | | | | |
| Latissimus dorsi | | | | | | |
| Pectoralis major | | | | | | |
| clavicular head | | | | | | |
| sternocostal head | | | | | | |
| Levator scapulae | | | | | | |
| Rhomboids | | | | | | |
| Serratus anterior | | | | | | |
| Trapezius | | | | | | |
| upper part | | | | | | |
| middle part | | | | | | |
| lower part | | | | | | |
| Subclavius | | | | | | |

FIGURE 61. Schematic representation of muscle activity in arm's forward upswing.

movement. In fact, active muscles antagonistic to movement, extend a steadying and stabilizing influence to the condyle. Both co-ordination patterns in Figures 60 and 61 concern voluntary movements beginning with a limply hanging arm. However, if abduction is performed by supinating the hand, this altered hand position influences co-ordination in shoulder musculature. With a supinated hand, the biceps brachii is active and its long head will then collaborate in the arm's abduction. This is only an example of how co-ordination in a joint's musculature is not only dependent on the joint's movement but even on the position and movement of adjacent joints. Such conditions make it so difficult to comment *a priori* on co-ordination in a special movement, e.g. a work move-

129

ment, without electromyographic registration, even if the coordination pattern is known from the arm's two basic movements.

The shoulder's muscles can be divided into three groups. The deep group consists of four muscles: the subscapularis, supraspinatus, infraspinatus, and teres minor. These muscles are inserted on the greater and lesser tuberosity of the humerus and into the capsule of the shoulder joint. The part of the muscle group whose direction of pull is the same as the actual direction of movement acts to facilitate movement. In abductive movement of the arm, the entire supraspinatus muscle acts as an abductor from the very beginning, just like the acromial part of the deltoid muscle. Only later in the movement do other and deeper muscles, such as the anterior and posterior part of the deltoid muscle, begin to act abductively. In the first stage of the movement, the latter muscles actively impede movement, but the head of the humerus is stabilized as a result of this antagonistic effort, dislocation otherwise occurring. At the beginning of the arm's forward swing the subscapularis muscle is an important muscle of movement, but the subscapularis, like other deep muscles, acts primarily as a stabilizer of the head of the humerus and is a moderator of the head's sliding movement across the surface of the joint socket.

The most important muscles of movement are the deltoid and the clavicular head of the pectoralis major. They are the most superficial shoulder muscles and are the ones with the most distal attachment. They provide the most favourable levers for movement, as the centre of arm movements is in the head of the humerus. But the centre of the forces which stabilize the condyle in the socket is in the vicinity of the humeral insertions of the deltoid and pectoralis major muscles. The closer the insertion of these stabilizing muscles to the head of the humerus, the longer the lever arms.

Just as stabilizing muscles in certain phases of movement act to further movement, muscles of movement may also act as stabilizers in certain phases. Since all parts of the deltoid muscle are active throughout abduction movement, both the anterior and posterior part of the muscle act as stabilizers in the first stage of the movement, i.e. as long as their respective resultant forces are below the sagittal axis of movement.

The muscles of the shoulder joint can be compared to a group of people raising a flag-pole. The people whose task it is to hold fast the pole's lower end and make sure it does not slide across the ground when the pole is raised hold on as close as possible to the bottom end of the pole, while those persons who are to raise the pole apply their forces as close as possible to the upper end of the pole.

The middle and third group, the sternal part of the pectoralis major, the latissimus dorsi and teres major are attached to the humerus in the area between the deep and superficial muscles' insertion sites. These muscles act both as stabilizers and as muscles of movement. These muscles provide, so to speak, a reserve for the other groups.

An analysis here of the mechanical task of every muscle in relation to its degree of activity would require a far too extensive discussion. But it may be mentioned that the reason why activity in a muscle does not increase continuously until a movement has been completed is because mechanical conditions are gradually changed during the course of a movement and because a muscle's ability to exert force varies with its degree of extension. Thus, activity in, e.g., both primary muscles of movement, the deltoid and pectoralis major, reach their activity maxima when the arm is abducted and flexed respectively, approximately 120° even though the arm's moment of inertia is greatest when the arm is raised 90°. This is mainly because the ability of muscle fibres to develop power declines the more the fibres are shortened. Therefore, a larger number of fibres, i.e. motor-units, must be engaged when the arm is raised 120° than when it is raised 90°, even if the moment of inertia is greater in the latter case.

### THE ELBOW

The elbow's flexors are the brachialis, biceps brachii and the brachioradialis, and its extensor is the triceps brachii. Lower arm muscles which arise from the epicondyles of the humerus and which pass the elbow are of completely subordinate importance in the flexion and extension of the elbow, and they are also inactive in its movements as long as there is no simultaneous movement in the carpal joints or fingers.

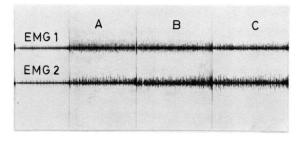

FIGURE 62. Activity in the biceps brachii (EMG 1) and brachioradialis (EMG 2) during voluntary bending of the elbow-joint when the hand is: A supinated, B semi-pronated, and C pronated.

The brachialis is the primary elbow flexor, Figure 62. The activity of other flexors varies with the position of the hand, i.e. with the hand's pronation and supination positions. In a semi-pronated position, activity is relatively evenly distributed amongst the three muscles. The more the hand is supinated when the elbow is bent, the more strongly the biceps brachii is engaged. However, if the hand is pronated activity of the biceps in elbow flexion is reduced. Alternatively, activity in the brachioradialis is somewhat greater when the hand is pronated than when it is supinated.

If the position of the hand is more decisive in activation of the biceps brachii than of the brachioradialis, then movement speed is primarily reflected in the degree of activity in the brachioradialis. The brachioradialis is the muscle of rapid flexion.

The triceps brachii acts antagonistically towards these three flexors. The medial portion of the muscle's three portions appears to be the primary elbow extensor. When greater power is required, the lateral and scapular heads are recruited. With slow flexion and extension of the elbow, there is good agreement between the muscle's anatomical prerequisites for achieving movement and its activation. With rapid flexion and extension, however, flexors are activated as early as at the end of the extension movement and the triceps brachii at the end of the flexion, Figure 63. This activity in the movement's antagonistic muscles may be interpreted as protection for the joint against mechanical loads which are too great.

The position of the pronation and supination axis in relation to direction of pull in the muscles of the lower arm provides most of the muscle with mechanical qualification to perform pronation and supination movements. However, electromyographic studies have shown that only a few muscles actively take part in the movements. Irrespective of the position of the elbow, i.e. the degree of elbow flexion, it is primarily the pronator quadratus muscle which is engaged in pronation. In most cases the pronator teres also takes part, particularly if pronation is performed rapidly or against resistance. But such muscles as the flexor carpi radialis, brachioradialis, and extensor carpi ulnaris do not appear to have any pronation function. It is primarily the supinator muscle which is engaged in the hand's supination movements. If supination is performed with a somewhat flexed elbow, even the biceps brachii is engaged. In all supination movements requiring great effort, the supinator muscle and biceps brachii form a functional synergist group. In both pronation and supination movements, there is reciprocal inhibition, whether the movements are made quickly or slowly.

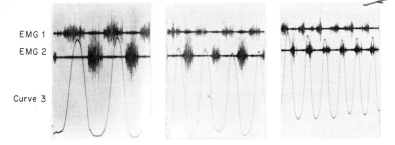

EMG 1
EMG 2

Curve 3

FIGURE 63. Activity in the biceps brachii (EMG 1) and triceps brachii (EMG 2) during voluntary bending and stretching of the elbow with different velocities.

Curve 3: electrogoniograph—rising curve indicates bending, falling curve indicates stretching.

### THE WRIST

The hand's dorsal and palmar flexion occurs around a radio-ulnar axis of movement which passes approximately through the centre of the head of the capitate bone. Perpendicular to this axis and also passing through the head of the capitate bone is the axis around which radial and ulnar abduction of the hand is performed. In isolated movements around these movement axes, i.e. without any simultaneous movements in the fingers, there is good agreement between the physiological activity of the muscles and their anatomical arrangement. However, one exception is the extensor carpi ulnaris, which is active in both dorsal and palmar flexion, Figure 64.

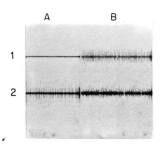

FIGURE 64.

EMG 1: Flexor carpi ulnaris.
EMG 2: Extensor carpi ulnaris.
A: Extension of the wrist.
B: Flexion of the wrist.

The hand's dorsal flexion is achieved by the extensor carpi radialis, extensor carpi ulnaris and extensor digitorum together. The flexors remain passive even with heavily forced dorsal flexion. In the hand's palmar flexion, the entire flexor group works synchronously with the exception of the flexor digitorum profundus, which generally remains passive in this movement. Alternatively, the extensor carpi ulnaris is active. Thus, it would appear as if this latter muscle supplements skeletal and ligamentous control of the ulnar part of the carpus, control being poorer in this part of the joint than in the radial part.

In abduction movements, there is reciprocal innervation between the extensor and flexor carpi radialis and between the extensor pollicis brevis and abductor pollicis longus on the one hand and extensor and flexor carpi ulnaris on the other hand. There are large individual variations regarding finger flexor and extensor muscles passing the wrist and whose direction of pull passes through or close by the axis of movement. In certain persons all these muscles remain passive; in others they are active in both ulnar and radial abduction and in still others, some muscle is active in one movement and passive in others, etc. With simultaneous EMG from the ulnar and radial part of the extensor digitorum, one finds that the muscle works as a unit. There does not appear to be any division of the muscle into portions with various functions, which is hardly what one might have expected to find in view of the muscle attachment's distribution among the four ulnar fingers.

## THE FINGERS

The anatomy of the hand, like its function, is unique in the animal kingdom. The form and function of individual joints and individual muscles are relatively well-known, but the function of the hand as a whole and its specific movements and characteristics have scarcely attracted the interest of scientists to any extent. And yet it is the total movements of the hand, the sum of sub-movements, which make the hand the eminent gripping organ that it is. It is also the hand's function rather than its anatomy which one tries to restore in treatment of an injured hand.

The thumb's opposition to the other fingers is a special characteristic of the human hand. This circumstance, plus the 35-odd muscles, many of which are built up of small motor-units capable of well-defined contractions, acting on the hand across about 20 joints, provides the basis for the truly varied pattern of movement of which the hand is capable. The extremely well-developed skin sensitivity on the hand's palmar side and in the fingertips is also a contributory factor and most essential to the many subtle and versatile movements of the hand. One can indicate the site for a stimulus on the hand with greater certainty than on the arm or leg. Two simultaneous impulses, e.g. pricks made with a pair of compasses feel like two separate pricks if the distance between compass tips is 3–8 mm for the fingertips, 1 cm on the palm, and 4–7 cm on the arm, leg, or trunk.

The thumb is undoubtedly the most important of the five fingers. In the rest position, its metacarpal bone forms an angle of 45–50° with the other metacarpal bones. The thumb's carpo-metacarpal joint is the only saddle joint in the body and a very stable one at that, despite the absence of ligaments. In the rest position, one of the joint axes is parallel to the axes in the thumb's interphalangeal and metacarpo-phalangeal joints. Opposition and reposition occur around this axis, a movement which may be regarded as modified extension and flexion. In this movement, there is compulsory rotation around the meta-carpal bone's longitudinal axis, enabling the thumb to move in or out from the palm and the palmar surfaces of the other fingers. Perpendicular to this 'opposition–reposition axis' is the 'adduction–abduction axis' around which the thumb is moved in and out from the index finger. A wide capsule and deformable cartilage make it possible for the thumb's carpo-metacarpal joint to be functionally designated as a ball-and-socket joint. The interphalangeal and metacarpo-phalangeal joints are connected in as much as movements in these joints are either performed in the same direction or one joint is kept still while the other is flexed or extended. But in normal conditions, one joint is never flexed while the other is extended or vice versa. Movement in the thumb's carpo-metacarpal joint can, however, be made independently of the movement and positions of the other two thumb joints.

When a large, cylindrical object is grasped with the hand, the thumb is opposed in the carpo-metacarpal joint and is flexed in both distal joints at the same time as the other fingers are bent. The opposite movement, when the thumb is repositioned and the other fingers extended, is made when the hand is opened. If a small object is grasped and held in a thumb–forefinger grip, there is even an adductive movement in the carpo-metacarpal joint, in addition to the aforementioned opposition and flexion movement, as part of the thumb's movements. This adduction becomes successively less pronounced as the thumb is opposed to the middle finger, ring finger, and little finger.

No fewer than eight muscles are available for thumb movement. These muscles stream in from different directions towards the thumb and make movement of the thumb possible in different directions.

The listing below shows which thumb muscles are engaged in different movements according to the relatively few electro-myographic studies made of this musculature.

In opposition (often in combination with flexion in the thumb's metacarpo-phalangeal joint and interphalangeal joint):

opponens pollicis[1]
abductor pollicis brevis[1]

flexor pollicis brevis[1]
adductor pollicis[1]
flexor pollicis longus[2]
abductor pollicis longus[2]

[1]Activity increases in strength successively when the thumb's position of opposition is moved, finger by finger, from the forefinger to the little finger.

[2]Active in position of opposition towards ring and little finger and, in certain cases, even towards the middle finger.

In reposition of the thumb to its normal position (often in conjunction with extension of the thumb's metacarpophalangeal and interphalangeal joint):

abductor pollicis longus
extensor pollicis brevis
first dorsal interosseus
extensor pollicis longus
opponens pollicis
flexor pollicis longus

In abduction:

opponens pollicis
extensor pollicis brevis
abductor pollicis brevis
abductor pollicis longus
flexor pollicis brevis
flexor pollicis longus
extensor pollicis longus

In adduction:

adductor pollicis
flexor pollicis brevis
first dorsal interosseus
extensor pollicis longus
flexor pollicis longus
opponens pollicis

As may be seen from the above, there are several thumb muscles which are in action simultaneously, both facilitating movement and antagonistic to movement. These latter muscles have a stabilizing effect on movement and on the joints. By displacement of total muscle activity in one direction or another so that at the same time as activity increases somewhat

137

in a muscle, it declines to the corresponding extent in another muscle, the thumb's direction of movement is changed. Thus, a somewhat altered direction of movement need not mean that one or a pair of muscles remain quite passive and new muscles are brought into action. In adduction of the thumb, it is not only the adductor pollicis which is active, just as the thumb's abduction is not only performed with the two abductors. These muscles are, indeed, especially active in the thumb's adduction and abduction respectively, but at the same time a number of other muscles are active as well, even if their anatomical names fail to suggest this. From the functional point of view, the anatomical names of the various thumb and thenar muscles are, therefore, somewhat misleading.

If the handle of a cup is taken with a thumb–forefinger grip, one finds that the flexor pollicis brevis is much more active than if the cup itself, or a glass as in Figure 65, is grasped with an 'open' thumb–forefinger grip. Just the opposite is true with activity in the opponens pollicis and abductor pollicis brevis. The activity of these muscles is definitely stronger in the latter than in the former case. The more the thumb is abducted, the less the flexor pollicis brevis is engaged in order to produce a fixed grip. On the other hand, the short thumb flexor appears to be the thenar muscle which plays the decisive role when one wishes to grasp a small object such as those ordinarily held in

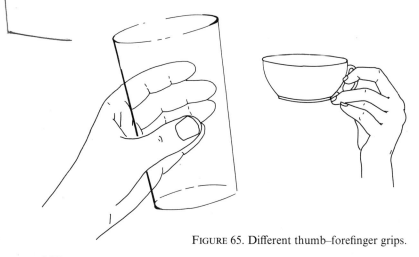

FIGURE 65. Different thumb–forefinger grips.

a thumb–forefinger grip, e.g. pens, pins, instrument handles, etc.

As mentioned previously, the four ulnar fingers are somewhat flexed when the arm hangs limply in the rest position. There is then no activity in the musculature of the hand and lower arm. The flexion position is because the passive forces on the fingers' palmar sides, primarily in the flexor digitorum profundus muscle, are greater than the extensive passive forces on the dorsal side. To a certain extent the degree of flexion is dependent on the position of the wrist. If the hand and lower arm rest on a table with the ulnar side down, one finds that the more the wrist is dorsally flexed, the more the fingers are flexed because of increased passive tension in the deep finger flexor muscle which develops when the muscle is stretched in dorsal wrist flexion. If the wrist is in palmar flexion, this causes a reduction of tension in the flexor muscle at the same time as tension on the extensor side increases. The flexion position of the fingers becomes less pronounced.

Actively straightening the fingers from the rest position involves an extension in the metacarpo-phalangeal joint and the proximal and distal interphalangeal joint. One could also speak of finger extension in extension of the interphalangeal joints alone with retained or increased flexion in the meta-carpo-phalangeal joint. In both cases there is polyarticular or diarthric movements. In extension and flexion, both interphalangeal joints are normally co-ordinated, i.e. if one joint is extended, so is the other and if one joint is flexed, the other interphalangeal joint is also flexed. However, there is no such coupling between the movements of the interphalangeal joints and the movements of the metacarpo-phalangeal joint in normal anatomical conditions. Thus, movements can be made in the metacarpo-phalangeal joint independently of the position and movement of the interphalangeal joints. And movement in the interphalangeal joints can be made independently of the position and movement in the metacarpo-phalangeal joints. The explanation for the coupling of the interphalangeal joints is to be found in the arrangement of the connective fibres in the finger's extensor expansion. Figure 66 shows schematically the arrangement of different fibre bundles in the extensor expansion. The extensor tendon, passing over the middle of

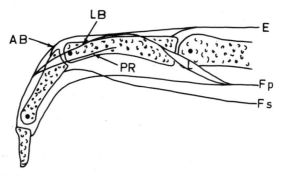

FIGURE 66. The extensor expansion.

E: Extensor digitorum; AB: axial branch; LB: lateral branch; PR: palmar reins; L: lumbricales; Fp: flexor digitorum profundus; Fs: flexor digitorum superficialis.

the metacarpo-phalangeal joint, whose deep layers are attached to the base of the basal phalanx, divides itself somewhat distal to the middle of the proximal phalanx into an axial and two lateral branches. The axial branch is attached to the base of the middle phalanx. Both lateral branches pass dorsolateral to the proximal interphalangeal joint and then join together towards their insertion at the base of the distal phalanx. From the lateral parts of the proximal phalanx, paths arise which run in a distal–dorsal direction towards the middle of the middle phalanx where they join the lateral branches. These paths, which are usually called the palmar reins, pass palmar to the movement axis of the proximal interphalangeal joint. Lumbricales and interossei, whose attachment tendons unite with the aponeurosis on a level with the proximal phalanx, join the dorsal aponeurosis.

As may be seen from Figure 66, the extensor digitorum is mechanically able to extend both the metacarpo-phalangeal joint and both the interphalangeal joints. An extension of all joints means that passive resistance on the flexor side, primarily the elastic resistance in the flexor digitorum profundus, must be overcome. However, this resistance cannot be overcome by the extensor digitorum alone in order to produce extension in all joints. A contraction of the extensor digitorum alone leads to a claw position, i.e. an extension of the metacarpo-phalangeal joint while the interphalangeal joints are

140

flexed. That this position develops appears to be because the extensor muscle's torque in the metacarpo-phalangeal joint is greater than the torque of the passive resistance on the flexor side, while the torque of passive resistance on the palmar side around the movement axes of the interphalangeal joints is greater than the extensive torque produced by the extensor digitorum on the dorsal side of these joints. A simultaneous extension of both the metacarpo-phalangeal joint and the interphalangeal joints requires an activation of yet another muscle. Electromyographic studies have also shown that the hand's lumbricales, in addition to the extensor digitorum, are also active in such polyarticular extension. The lumbricales arise from the tendon of the deep finger flexor muscle radial to the proximal part of the metacarpal bone and are attached to the finger's extensor expansion on a level with the proximal phalanx. When the lumbricales contract they accordingly pull the tendon of the deep finger flexor distally at the same time as they pull the distal part of the finger's extensor expansion proximally. By pulling the tendon of the deep flexor muscle distally, the part of the tendon which is distal to the origin of the lumbricales, i.e. the part of the tendon which is between this origin and the flexor tendon's insertion at the distal phalanx, is unloaded. Thus, the active lumbricalis reduces passive resistance on the flexor side of the interphalangeal joints at the same time as the muscle produces relative extension on the dorsal side of the joints. Since the lumbricalis pulls distally on the flexor tendon with the same force as on the extensor expansion proximally, there is no force which can flex the metacarpo-phalangeal joint, even though the muscle passes palmar to the movement axis of this joint. It should be pointed out here that lumbricales and the flexor digitorum profundus are never active at the same time. The ability of the lumbricales to reduce the power of flexion of the interphalangeal joints and contribute to extension of the interphalangeal joints without affecting the metacarpo-phalangeal joints is unique and differs in this respect from the interossei. Total finger extension is, thus, controlled by the active extensor digitorum, active lumbricales and the passive flexor digitorum profundus.

If the fingers adopt a claw position, i.e. with an extended

metacarpo-phalangeal joint and maximum flexion of the interphalangeal joints, movements in the metacarpo-phalangeal joint, i.e. flexion and extension, are only performed by an interplay between the extensor digitorum and flexor digitorum profundus (possibly the flexor digitorum superficialis as well) but without active participation by the hand's own muscles.

An extension of the interphalangeal joints alone with maintained flexion or increased flexion in the metacarpo-phalangeal joint requires great power and co-operation of the interossei, whose most important function is otherwise to abduct and adduct the fingers at the metacarpo-phalangeal joints. Electromyographic studies have shown that both interossei and lumbricales are strongly active in this position with flexed metacarpo-phalangeal joints and extended interphalangeal joints. Relatively weak activity also occurs in the extensor digitorum. Extension of the interphalangeal joints is produced by the three muscles in common, while flexion in the metacarpo-phalangeal joints is the work of the interossei.

In the type of finger bending occurring when the hand is closed, both the flexor digitorum profundus and extensor digitorum are active, according to electromyographic registrations, both when the hand is closed from the rest position as well as when the extended hand is closed.

It has previously been pointed out that isolated contraction of the extensor digitorum leads to a claw position because of passive resistance in the flexor digitorum profundus particularly. It seems therefore odd that increased resistance in the flexor digitorum profundus, which an activation of the muscle must imply, leads to complete finger bending, i.e. flexion in both the interphalangeal joints and metacarpo-phalangeal joint, and not to a claw position. There is no really satisfactory and exhaustive explanation for this. However, it is easy to note that if a claw position is voluntarily adopted, the extensor digitorum is much more active than when one voluntarily closes the hand, Figure 67. Thus, it is possible that it is the degree of activity in the extensor muscle and, thus, even the magnitude of the extensive torque around the metacarpo-phalangeal joint which determines if this joint is extended or flexed when flexion and extension forces act simultaneously. However, it is not uncommon to find that activity in the pal-

A          B

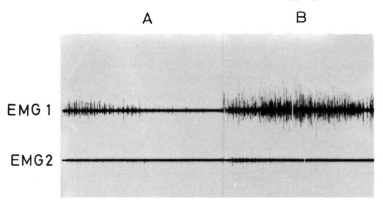

EMG 1

EMG 2

FIGURE 67.
A: Ordinary closing of the hand until the fingertips lightly touch the palm.
B: Voluntary claw position of the hand.
EMG 1: Extensor digitorum; EMG 2: second dorsal interosseus.

mar interossei is registered in finger bending, and these muscles then work towards flexion in the metacarpo-phalangeal joints. If the fingers are abducted somewhat, as they often are when the fingers are extended, it is possible that the palmar interossei are activated in their capacity as the fingers' adductors. At the same time as they then adduct the fingers, they aid in flexing the metacarpo-phalangeal joints. Their extensive influence on the interphalangeal joint is apparently outweighed by the active flexor digitorum profundus.

When the finger is flexed and the flexor muscle shortens, the extensor expansion must lengthen and, thus, the axial branch and the lateral branches are displaced distally. This lengthening of the extensor expansion takes place under the control of the active extensor digitorum. Distal displacement of the axial branch immediately results in a flexion in the proximal interphalangeal joint. This flexion also causes the lateral branches to relax. The distance between branches and the movement axis of the proximal interphalangeal joint declines, and the more the proximal interphalangeal joint is flexed, the more the branches relax.

These relaxed lateral branches now permit flexion of the distal interphalangeal joint, flexion which can continue until the lateral branches are tensioned once again.

143

The aforementioned palmar reins also play an essential role in the regulation of movements of the interphalangeal joints. Active flexion of the distal interphalangeal joint cannot be made without extension of the palmar reins, and this extension of the reins automatically carries with it flexion around the proximal interphalangeal joint, as the reins pass palmar to the joint's movement axis. The force is provided by the deep flexor muscle and this is transferred to the palmar reins via the distal phalanx and lateral branches. In normal conditions it follows that the proximal interphalangeal joint must be flexed at the same time as the distal joint. When the interphalangeal joints are flexed to a maximum, the palmar reins are tensed and, accordingly, prevent any extension of the proximal interphalangeal joint but facilitate an extension of the distal interphalangeal joint, as the reins pass into the tendon insertion on the terminal phalanx. An extension of the distal joint causes relaxation of the palmar reins, also making extension of the proximal joint possible.

There are persons who can hyper-extend the proximal interphalangeal joint at the same time as they flex the distal interphalangeal joint. This appears to be because the palmar reins are dorsally displaced to the axis of movement of the proximal interphalangeal joint; they are then tensed and lock the proximal joint in an extended position when the distal joint is flexed.

If one flexes a separate finger in the proximal interphalangeal joint, which most people can do voluntarily, one finds that this finger bending can be performed without affecting the other fingers and without flexion in the distal interphalangeal joint, and the distal phalanx feels as if it were 'hanging loosely'. These conditions are indications that finger flexion was performed with the flexor digitorum superficialis while the deep finger flexor remained passive and that the palmar reins are relaxed. Thus, such finger bending deviates markedly from the habitual finger bending previously described.

In the habitual finger bending occurring when the hand is closed and all four fingers flex simultaneously, the flexor profundus is the primary finger flexor. The superficial finger flexor muscle then remains completely passive or may show weak activity in certain cases. The degree of this activity appears to be related to the position of the wrist. If the wrist is extended, there is usually no activity, but if the wrist is flexed and length-tension conditions for the deep finger flexors become unfavourable, the superficial finger flexor is also activated. Collaboration is then required of both finger flexors so that the hand can be closed. The flexor digitorum superficialis is what one might call a contingency muscle, prepared to intervene when the flexor digitorum profundus is unable to produce the requisite force alone.

In addition to the common finger extensor, the forefinger and little finger each have their own extensor, making more powerful extension possible. No functional difference between these separate extensors and the common extensor muscle has been demonstrated. However, if the forefinger is extended for a sufficiently long time, one can observe a certain decline in activity in the extensor digitorum at the same time as there is increased activity in the extensor indicis.

The four dorsal interossei are responsible for abduction of the second and fourth fingers. The fifth finger has its own abductor, the abductor digiti minimi. The three palmar interossei adduct the forefinger, ring and little fingers.

Abduction and adduction movements can be performed freely when the metacarpo-phalangeal joints are extended. But if these joints are flexed, the fingers are automatically adducted, even without the active participation of the palmar interossei, and the ability to abduct disappears successively. This limitation in movement is because the collateral bands in the metacarpo-phalangeal joint are tensioned as the joint is flexed. If a spherical object is grasped and significant hand strength is required to hold onto the object, this ligamentous locking of the metacarpo-phalangeal joints greatly facilitates the task of the palmar interossei: to fix the grip and prevent the fingers from gliding apart.

A finger cannot be rotated voluntarily around its longitudinal axis. But the fingers are often exposed to rotative external forces. If an object is held in the hand as in Figure 68 and an attempt is made to twist this object in a direction more or less perpendicular to the fingers' longitudinal axes, forces develop which act to rotate the fingers. These rotative forces are counteracted by the joint capsules and ligaments and even by active interossei.

145

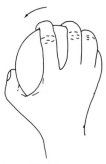

FIGURE 68. When twisting a lid in the direction of the arrow the first and second dorsal interossei will be active to counteract the forces which act to rotate the fingers around their longitudinal axis.

### THE GRIP

The gripping movements we continuously perform with the hand in our daily activities are infrequently isolated flexions and extensions, abductions and adductions, or opposition and reposition movements, but a combination of all these movements. This combination can be varied in many ways. As a rule, muscle co-ordination is also more complicated in these gripping movements than they are in free finger movements. However, with knowledge of the muscles' engagement in free movements, it is possible for one to acquire a relatively clear picture of muscle co-ordination in a gripping movement from the position of the fingers and wrist and from the power of the grip. Despite the complexity of the co-ordination pattern, it is still difficult to get an overall picture of the multitude of hand movements and grips the hand produces in its different functions.

Isolated movements in the various joints of the hand and fingers, like the functions of the various muscles in these movements, have been rather extensively analysed. However, it is more difficult to gain a clear picture of the movements which the hand as a whole performs. One generally speaks of opening and closing the hand, of gripping an object, of holding something in the hand, etc. But these phrases say nothing about how the movements are performed.

For purely practical reasons, attempts have been made to

analyse the function of the hand *inter alia* in order to evaluate the degree of uselessness of an injured hand. Thus, the function of the hand has been divided up according to the grip required by an object's shape. There is a cylinder grip, ball grip, ring grip, pincer grip, etc. But this is actually a division into functional final results and says nothing about the hand's movements or movement potential. Nor is there any anatomical or physiological basis for the three classes into which some authors divide the function of the hand. They speak of 'gripping with whole hand', 'gripping with thumb and finger', and 'gripping with finger and palm'.

Sometimes one speaks of the hand's 'normal grip' with which the most powerful hand grip can be achieved. With the thumb in opposition, one grasps the object with the thumb from one direction and with the four ulnar fingers from exactly the opposite direction. If the thumb is adducted to the ring finger and the object is held between the four ulnar fingers and carpus, one speaks of a 'thumbless grip'. A bag, for example, is carried with an 'open grip'. The metacarpo-phalangeal joints are extended while the interphalangeal joints are flexed, the bag handle resting on the middle phalanges. If the object is held between the thumb's fingertip and one of the other fingertips, most often the forefinger, the object is then held in a 'fingertip grip'. The strength correlation between normal grip and thumbless grip is 0·58, between normal grip and open grip 0·49 and normal grip and fingertip grip 0·24. Common to most classifications is the fact that the greater part of the hand's movements have been excluded.

One system which is based on both anatomical and physiological conditions and which appears to possess more general applicability has been devised by the English worker Napier. He speaks of two main groups of hand movements: grip and non-grip movements. In the latter, the hand is shaped into a supporting organ as, e.g., when you push an object in front of you or when the hand is shaped into a 'hook upon which a bag is hung'. In these movements, the hand remains relatively passive in relation to both the object and the arm. Even if these 'non-grip movements' play a vital role in daily life, it is the true grip movement—the prehensile movement—which is of greatest importance to use from the kinesiological point of view.

In his analyses of the hand's functional movements, Napier found that the multitude of grip movements performed by the hand are variations of two fundamental but widely differing types of movement. In any gripping movement, it is essential that the grip on the object is really steady enough to permit desired manipulation of the object. This stability can be achieved in two ways. The object can be held between the four, more or less bent ulnar fingers and palm, and this grip is fixed with the thumb which is held on a plane with the palm, Figure 69. This grip is called the *power grip*. The object can also be held between the thumb and palmar side of the other fingers.

FIGURE 69. The hand's power grip.

This grip is called the *precision grip*, Figure 70. Both these grips appear to cover the entire range of the hand's gripping potential.

Both grips are distinctly different anatomically and physiologically. It is not an object's shape which primarily determines which grip is used in one situation or another but the act to be performed with the object (compare the grip on a tennis racket and on a golf club). Let us say one wishes to remove the lid of a jar. First one applies a power grip and loosens the lid with this grip. As soon as initial resistance has been overcome, one goes over to a precision grip and completes the act with it, finally lifting off the lid with the same grip. The aforementioned cylinder grip, named after the object's shape, is a precision grip if the 'cylinder' is a pen or a jeweller's hammer but a power grip if the 'cylinder' is an ordinary hammer handle or trowel.

In certain cases the size of the object may be decisive in the choice of grip. A larger object can be held with a precision grip rather than with a power grip. A large spherical object, no larger than can be held with one hand, is held in a precision grip. The object is held between the tips of the heavily abducted fingers and thumb. Extremely small objects are also held in a

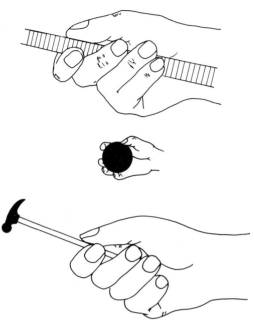

FIGURE 70. Different precision grips.

precision thumb-forefinger grip. The stability of these grips is really more of a sensory than mechanical matter. Even such features as the object's weight, temperature, degree of moisture, and consistency influence the grip. Even if an object's shape and nature may affect the shape of the grip in certain circumstances, it is the intended use of the object which ultimately determines the gripping pattern, i.e. if a power movement or precision movement is to be executed with the object. In most cases it is a question of *either/or*. But *both and* also occur.

In the power grip the thumb is adducted in the carpo-metacarpal joint and abducted in the precision grip. It should also be mentioned that a thumb in action is never in such a position that the carpo-metacarpal joint is in a neutral position, as the ligaments are then slack and gaps develop between joint surfaces. The thumb can only achieve a sufficiently stable grip in an abduction or adduction position. In the precision grip the object is held between the thumb and other palmar

149

surfaces of the fingers. In this grip the thumb is abducted and medially rotated in the carpo-metacarpal joint and also flexed in the metacarpo-phalangeal and interphalangeal joints. Skin sensitivity can be best exploited in this thumb position in order to provide the best means for fine adjustment of the grip.

The degree of precision required in a power grip is reflected in the position of the thumb. If little or no precision is required, the thumb is placed on the dorsal side of the fingers' terminal and middle phalanges, locking the grip on the object. The 'fist grip' is the most precision-free power grip, i.e. the extreme power grip. If a certain degree of precision is required in the power grip, the thumb is abducted, extended and rotated medially. The greater the degree of precision needed, the more these movements are accentuated.

In the power grip, the object lies between the four more or less flexed ulnar fingers and palm. The degree of finger bending and the area of the palm put into use depends on the object's size. In addition to being flexed, the fingers are rotated laterally somewhat and tilted towards the hand's ulnar side. In the precision grip, the fingers are flexed and abducted in the metacarpo-phalangeal joints, which subsequently contributes towards increasing the extent of the hand and the ability to rotate the fingers axially. The degree of flexion and axial rotation depends mainly on the object's size and shape. The smaller the object, the greater the demand for precision, and in order to satisfy this demand the grip is displaced towards the radial fingers and ultimately the thumb and forefinger, the fingers best suited to fine control.

The hand's position in relation to the forearm is also different in both grips. In the power grip the hand shows ulnar abduction but neither dorsal nor palmar flexion. In the precision grip there is dorsal flexion but neither radial nor ulnar abduction.

As is well-known, it is possible for one to indicate the position of a body part without actually seeing the body part in question. Similarly, one has a good idea of a movement's magnitude, direction, and speed without using sight. This orientation ability is possible thanks to the sensory organs: proprioceptors in muscles, tendons, ligaments, joint capsules, the labyrinth of the inner ear, etc. which respond to mechanical

conditions such as gravity, acceleration, speed, extension, and pressure. The learning of most of our hand and arm movements occurs under visual control. But we gradually learn to perform the movements without our eyes having to follow the movements; we perform the movements proprioceptively so to speak. With what degree of precision can one then perform such a proprioceptive movement as compared to visually controlled movement? The latter can be executed with almost perfect precision.

Certain subtle movements must always be performed under visual control, even if they are repeated continuously, e.g. the movements made by the hand when writing or drawing. The study of learnt and non-visually controlled movements has shown that precision is better if movements are made horizontally rather than vertically and that precision is better in the right hand and arm in right-handed people than in the left (and the reverse in left-handed people). It has also been shown that a movement's direction is characterized by greater accuracy than the movement's length. In general, movements become too long and display a tendency towards more lateral direction than intended. If, for example, the hand is to reach a point 60 cm in front of a test subject the movements are usually an average of 3–4 cm too long and show a lateral deviation of 2–3°.

It has been found that such movements made with the right hand and at a constant speed are made with greater precision if the movements take place obliquely forward and to the right. The opposite movement, i.e. from the front right in towards the body, is made with almost equal precision. Precision is poorest in movements straight forwards and back again or obliquely forward left and back. Movements made perpendicular to the underarm's longitudinal direction are always performed with greater certainty than movements in the longitudinal direction of the underarm.

To get an idea of how accurately a hand movement can reproduce a fixed distance without visual control of the movement, subjects were shown a picture upon which straight lines with lengths from 0·6–40 cm were drawn. With their eyes fixed on the indicated line in the picture, subjects were asked to move a slider the same distance as the line in question. As a rule, the short lines were reproduced with movements which were too long and the long lines with movements which were too short. Vertical movements were an exception, as vertical descending movements were consistently too long.

The accuracy with which one can repeat a certain hand or arm movement also depends on the frequency of execution. If the frequency is between 20–40 movements per minute, accuracy remains about the same. However, if frequency increases still further, accuracy declines. Movements then become more ballistic and control over the movements is especially impaired when the direction of movement is changed. On the other hand, precision of movement declines with intervals of rest; the longer the pauses, the poorer the accuracy. Thus, the greatest precision is achieved when the hand is moved back and forth on a forward-lateral course with a frequency of about 30 movements per minute.

# 7 Kinesiological Analyses in Sports and Work

As already mentioned in Chapter 2 the primary purpose of any kinesiological analysis is to acquire the most accurate and objective view possible of how a movement is performed in detail and how it is produced. Using existing knowledge regarding man's anatomical and physiological pre-requisites and limitations, it is then necessary to determine if the movement was performed in the optimum manner. Was it performed in the most functional manner? Could it have been made more effectively still? Or could it have been performed in some other, equally effective and perhaps less exhausting way? Does the movement really require so much force or does it subject the organism to such heavy or unfavourable loads that there is considerable risk of causing some loading injury if the movement is repeated? These are some of the questions which kinesiological analyses seek to answer.

Since there is no limit to the multitude of human movements and the purposes of movements vary, kinesiological analyses are not always able to follow the same programme. Several examples will serve to illustrate here how the author and co-workers sought to solve different kinesiological problems. The methods used by other workers can be studied *inter alia* in the ' collection of reports on kinesiological studies submitted to the two I.C.S.P.E. congresses in Zürich (1967) and Eindhoven (1969).

## THE GOLF SWING

In an analysis of the golf swing, a movement in which all of the larger body segments are engaged, the body's supporting

153

F

forces, the muscular activity in about twenty muscles, and the analytically necessary times for elements in the movement cycle (e.g. the start of the swing and when the club head meets the ball) were registered. The movement cycle was filmed synchronously with these registrations at 64 frames per second. Two force-plates, one for the right and one for the left foot, Figure 71 were used to measure the supporting forces. Figure 72 shows one of about 300 such registrations made. The analysis was based on the fact that all non-symmetrical body movements are reflected in reaction forces, forces of support, from the ground on which one stands. Symmetrical movements here refer to bilateral movements by corresponding parts of the body on the right and left sides which are performed at the same time, with the same speed, and in opposed directions. In a movement cycle such as the golf swing there are, however, no such symmetrical movements which can significantly affect the direction and magnitude of the forces of support.

Since the golf swing chiefly embraces rotatory movements, the kinetic equation $M = J.\ddot{\theta}$ is primarily applicable. It is consequently the movements of the larger parts of the body and the quicker movements which determine the magnitude of the reaction forces and of the moments of reaction. The body may here be divided into three mechanical segments which domi-

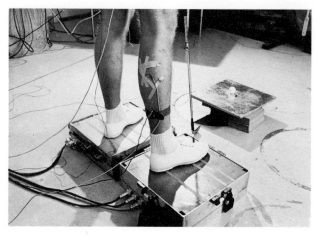

FIGURE 71. Force-plates for registration of reaction forces in performing a golf swing.

154

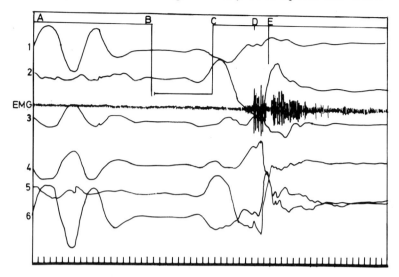

FIGURE 72. Recordings of forces of support and muscle activity.

The curves 1, 2, and 3 are the right foot's forces of support (Z, X, and Y). When the curve 1 (Z) sinks, the vertical pressure increases, when the curve 2 (X) sinks, there is a horizontal backward-directed force, and when the curve 3 (Y) sinks, there is a medially-directed force.

The curves 4, 5, and 6 are the left foot's forces of support (Y, X, and Z). When the curve 4 (Y) sinks, we have a medially-directed horizontal force, when the curve 5 (X) sinks, there is a forward-directed horizontal force, and when the curve 6 (Z) sinks, the vertical force decreases.

The comparisons are made with the conditions obtaining during the first part of the period B–C, the address. The backswing takes place during the period C–D, the down swing during the period D–E, and the impact is in E. EMG is from the oblique abdominal muscles of the right side. Time markers $\frac{1}{10}$ s.

(*See* further in the text.)

nate the movement and whose timing is decisive to the swing's mechanical results, i.e. to transfer as much as possible of the body's and club's kinetic energy to the ball.

The largest body segment is the trunk which rotates around an axis which runs approximately in the body's axis of symmetry through the lower part of the spine. Through a firm grip on the shaft of the club the hands and club form a mechanical unit, a 'hand–club' segment. This segment rotates round an axis which lies between the left and right wrist and

155

runs in a radio-ulnar direction. However, this axis of rotation is in turn moved round the axis of rotation of the trunk. Through the grip of the hands on the club the arms and shoulders are also coupled to form a mechanical unit, a 'shoulder–arm' segment, whose axis of rotation does indeed possess a somewhat lateral movement but on the whole lies within the upper part of the spine.

The horizontal movement for the entire system, i.e. the three segments together, may be regarded as a rotatory movement around a vertical axis which moves within a very limited area

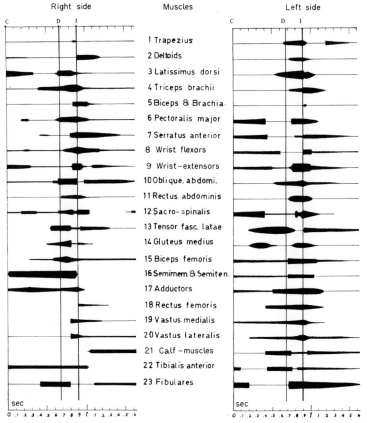

FIGURE 73. Muscular co-ordination in the golf swing.

The backswing takes place in period C–D, the downswing in D–I. The club meets the ball at I. After I comes the follow-through.

within the body's supporting surface. The other three axes may be considered to rotate around this axis. In case of equal load on the right and left foot, this axis runs midway between the right and left foot and a few centimetres in front of the malleoli. If, on the other hand, the load is greater on the right foot than on the left, it will lie nearer to the right foot than the left. Vertical movement for the trunk segment is so slight that its vertical reaction forces may be disregarded in comparison to the corresponding movement in the other two segments. The vertical movement in these segments may here be regarded as a rotation about a horizontal axis.

A collection of the electromyograms, Figure 73. The direction and magnitude of the forces of support and film sequences provide a detailed picture of the torque of the body segments and from these the kinetics and kinematics of the golf swing.

## THE WEIGHTLIFTER

A somewhat less complicated movement cycle than the golf swing is performed by a weightlifter when executing a press. With severe knee and hip flexion but a relatively straight back, the weightlifter bends forward and down and grasps the dumbbell with a pronated grip. The dumb-bell is hoisted to shoulder height with rapid arm flexion at the same time as the body is raised and the weightlifter adopts a standing position. He is now in the 'clean' phase, Figure 74. After a brief pause in this position, the dumb-bell is thrust vertically upwards. When the arms have been extended to a maximum and the body is in balance, the press has been completed. The entire movement occurs more or less in the sagittal plane. If the co-ordination pattern of a top weightlifter is compared to that of a less advanced lifter clear differences are noted, just as when one compares different skilled golf players. Figure 75 shows, for example, how one lifter holds the dumb-bell with a minimum of muscular effort at the 'clean' and then concentrates his muscular effort and power development in a rapid, almost explosive final press. The other lifter is unable to use his muscular forces economically. His musculature is engaged throughout the entire movement cycle and the final press is performed more slowly.

157

FIGURE 74. Weightlifter in the 'clean' stance in the press.

Arm extension in the final press and activation of the triceps brachii are always preceded by contraction of elbow flexors and minor but rapid bending of the elbow. The triceps brachii is accordingly stretched and acquires a favourable starting position for arm stretching. When the triceps is subsequently activated and the elbow extended, activity disappears in the biceps brachii and brachialis in a skilled weightlifter. However, in the other man the flexors remain active throughout the entire final press. Thus, the movement becomes slower and uneconomical. In this instance, there is co-contraction with both flexors and extensors in action. The triceps brachii must overcome the weight of the dumb-bell when it extends the elbow and even overcome the resistance presented by the active flexor muscles. This is, of course, a less than successful pattern of co-ordination.

158

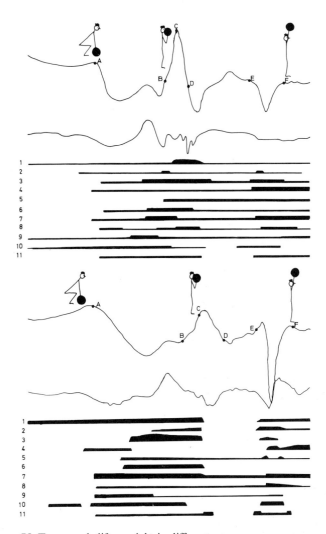

FIGURE 75. Two people lift a weight in different ways.

*Top*: a fairly good weightlifter; *bottom*: a well-trained weightlifter. The upper trace for each person shows the body's vertical force, and the trace drops when this force increases. The lower trace shows the horizontal force in a forward and backward direction, and the trace falls when this force is retrograde. Muscular activity registered electromyographically is shown schematically under the traces for each person.

A–C: the weightlifter bends down, grasps the dumb-bell and raises it to the 'clean' position. C–E: the 'clean' phase. E–F: the final press.

1: Lower arm extensors; 2: lower arm flexors; 3: biceps brachii and brachialis; 4: triceps brachii; 5: deltoid; 6: brachioradialis; 7: trapezius; 8: pectoralis major; 9: latissimus dorsi; 10: erector spinae; 11: rectus abdominis.

159

## THE ARCHER

In a laboratory study of muscular co-ordination in archery, an archer stood on two force-plates, one for the right foot and one for the left foot, Figure 76. Body balance was checked

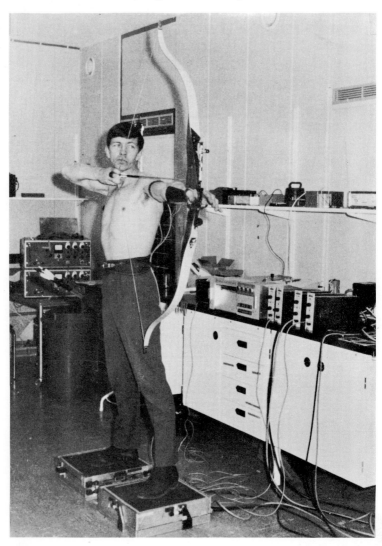

FIGURE 76. Force-plates for registration of body balance in archery.

with these plates. When the archer stood on the force plates in a comfortable position, load was usually evenly distributed on the right and left foot. As may be seen from curves 1 and 5 in Figure 77, body weight is displaced somewhat to the right at the draw. At the same time, load on the left foot is reduced to a corresponding degree. In the two curves in Figure 77, it should be pointed out that a rise in curve 5 means that pressure on the corresponding plate increases, while a rise in curve 1 implies reduced pressure on the corresponding plate. The nonsymmetrical load adopted during the draw is retained unchanged during aiming, but body balance is disturbed upon arrow release. Load on both feet is reduced for a brief instant and load is even shifted towards the right foot for a few tenths of a second before the body once again adopts a position with a more even distribution of body weight on both feet. These brief changes in load are undoubtedly recoil effects. The bilateral reduction in pressure on the plates may possibly seem

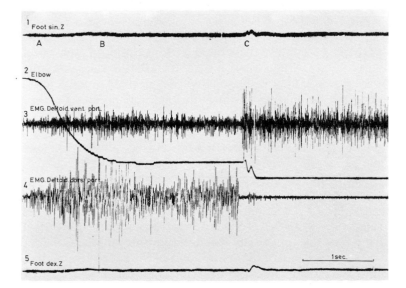

FIGURE 77. Archery.

Recording of the vertical pressure on the left foot, curve 1, and the right foot, curve 5. The muscle activity in right deltoid muscle, curves 3 and 4, and the flexion of the right elbow, curve 2.

Period A–B draw; period B–C aiming; release in C.

odd, but the reason is because the centre of gravity common to archer and bow drops. It has not been established whether or not this is because of knee bending, because the archer sinks the bow, or because of some other movement of brief duration.

Curve 2, Figure 77, shows how the right elbow is flexed during the draw (the curve falls). During aiming the archer holds the elbow very firmly fixed in a definite position. The joint is flexed further upon release of the arrow.

When the right arm's retrograde tensile force, i.e. the force applied to the bow string, suddenly terminates upon release of the arrow, the muscles connecting the left arm to the back and the scapula which are also in action during aiming, would fling the arm backwards if these muscles at arrow release failed to reduce their force or if their force were not counter-balanced by forces exerted in the opposite direction. This is also true of the muscles which keep the left elbow extended and the muscles which keep the left wrist extended. As may be seen from the table in Figure 78, there is a sudden and powerful increase in activity in the flexors of the left lower arms, i.e. the muscles which flex the wrist, and the flexors of the left elbow, i.e. the biceps, and in the left side, the pectoralis major, upon arrow release. Studies have also shown that at the same time as there is a sudden increase in activity in these muscles upon arrow release, activity declines just as suddenly in several of the antagonists of these muscles.

The rise in activity in the serratus anterior on the right side and the ventral part of the deltoid on this side, at the same time as the posterior part of this muscle is silenced, is surely due to the archer's attempt to fix the right shoulder girdle at the moment of arrow release and prevent jerky retrograde movement of the shoulder girdle which would otherwise be the result of a sudden termination of resistance from the bow string. The task of the serratus anterior is to pull the shoulder girdle forward and of the ventral part of the deltoid to conduct the arm forward while the dorsal part of the deltoid draws the arm back.

The reason why there is simultaneous increase in force in the muscles which extend and flex the right wrist, i.e. the extensors (no. 1 in the table) and the flexors (no. 4 in the table) of the lower arm, in conjunction with arrow release is probably

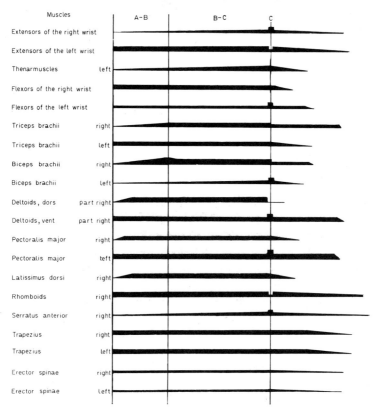

FIGURE 78. Schematic picture of the degree of muscle activity in archery. The stronger the line, the stronger the activity.

Period A–B draw; period B–C aiming; the release in C.
(*See* further in the text.)

because the archer has to increase the power with which the wrist is kept fixed. In the recorded increase in activity in both muscle groups, one is unable to determine the degree of activity in the individual muscles. The interplay between the fingers' flexors and extensors at the moment of arrow release would naturally be of interest, but such a detailed study would require registration of activity in each individual muscle. In that case wire electrodes would have to be used and not surface electrodes as were used in these experiments.

However, the study did show that undisturbed body balance

and a minutely co-ordinated interplay among many different muscles upon arrow release is of fundamental importance in archery.

### LIFTING IN A FACTORY

Insight into the human pattern of movement is of primary importance in the design of proper working conditions and working movements. Kinesiology has also acquired a prominent position in ergonomics.

Ergonomics is the scientific study of man and his work. The word ergonomics comes from the two Greek words *ergon* (work) and *nomos* (law). The ergonomist studies how different work steps and working conditions influence the human organism. He tries to design places of work, machines, work benches, etc. in accordance with human anatomy and physiology and to prevent unnecessary and unsuitable loading as much as possible.

In a crisp-bread factory, where work is largely carried out by machines, there is still some manual lifting involved in the manufacturing process. The baked bread pieces are transferred by conveyor belt from the ovens to a packaging department. At the end of this conveyor belt there is usually a bench at which a woman is placed to inspect the bread, pick out all the broken or deformed pieces and transfer fault-free pieces to another conveyor or to a trolley. Instead of picking up one piece at a time, which would be far too time-consuming, she usually collects a stack of 35–40 pieces of crisp bread and then transfers this stack to the other conveyor or to the trolley, Figure 79.

The technique used in this lift varies, depending on whether the lift is to the conveyor or trolley. In the former case, the stack of crisp bread is usually lifted from the 85 cm high bench to a conveyor belt 38–40 cm above the bench and about 40 cm behind the front edge of the bench at which the sorter stands. Thus, the lift movement is upward and forwards. The stack of bread is held in a diagonal grip, i.e. between the two palms facing each other. During the lift the right and left hands describe almost parallel courses of movement. This lift will henceforth be termed a *symmetrical lift*, as load is just about

FIGURE 79. Bread lifting in a factory.

evenly distributed on the right and left arm.

Using the other lift method, which is non-symmetrical, the bread stack is transferred from the aforementioned bench to a trolley behind the woman. She does admittedly pick up the bread stack with a diagonal grip but she then tips the bread stack onto her left arm and holds her right hand on top of the stack. The left arm could here be termed a lift arm and the right arm a support arm.

The kinesiological part of the ergonometric examination of the working conditions of these women consisted of an analysis and comparison of the different lifts. Since such an analysis for various reasons could not be performed when work was actually in progress in the factory, the examination was made in a kinesiological laboratory.

In order to reproduce a particular working movement adequately in the laboratory, it is necessary to build a replica of the particular place of work in the laboratory. As a rule, only those parts of the place of work which directly involve and influence the movement are required. Figure 80 shows how the place of work was designed in the laboratory so as to analyse crisp bread lifting. The bread is lifted here symmetrically from a 85 cm high and about 40 cm wide bench to another

165

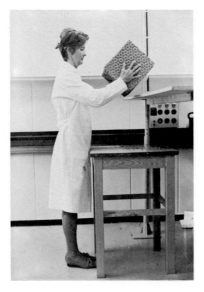

FIGURE 80. Bread lifting in the laboratory.

bench, 125 cm high, behind this bench. The latter bench, corresponding to the conveyor belt, was supplied with strain gauges which indicated when the bread stack was placed on the bench, i.e. when the lift movement itself had been completed. In these trials, the bread stack usually weighed 4–5 kg and contained 35–40 pieces.

In the study of the non-symmetrical lift, the bread was lifted from the aforementioned 85 cm high bench to a bench behind the test subject. The height of the latter bench, corresponding to the factory trolley, was 58 cm in certain trials and 153 cm in others, corresponding to the upper and lower levels in a fully packed trolley. Thus, both high and low non-symmetrical lifts were performed in these laboratory trials.

The most essential and important forces and torques included in these lifts are represented schematically in Figure 81. The muscles and the degree to which these muscles are engaged in the different torques and in the different lifts were electromyographically examined. The investigations showed that greater muscular engagement is required in order to stabilize the joints than to perform the movements themselves.

A special examination was made in order to study the female's lifting capacity in these special lifts, i.e. maximum

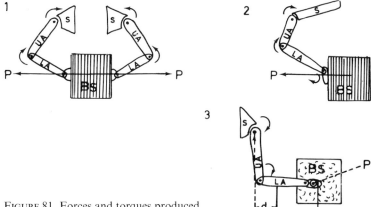

FIGURE 81. Forces and torques produced in lifting a bread stack.

1: In the horizontal plane.
2: In the frontal plane.
3: In the sagittal plane.
BS: Bread stack; S: scapula; UA: upper arm; LA: lower arm.

*P*: The hand-force ( = the reaction force) against the bread stack; *G*: the weight of the bread stack.

$d_1$: The horizontal distance between the axis of shoulder joint and the centre of mass of the lower arm; $d_2$: the horizontal distance between the axis of shoulder joint and the centre of the mass of the bread stack.

(*See* further in the text.)

lifting power. The lifting capacity of 35 women from 21–56 years was studied. A dummy load was used in this study instead of a stack of crisp bread. This dummy load consisted of a 35 cm long board to the ends of which two wooden discs about the size of a piece of crisp bread were affixed. As in the symmetrical bread lift, subjects held this dummy load in a diagonal grip, i.e. the palms were pressed against the wooden discs facing each other. With the dummy load in this grip subjects pressed the dummy load with maximum effort in a vertical direction upwards towards a horizontal 'handle' on a dynamometer. This dynamometer registered the force applied to the 'handle', i.e. the lifting power directed vertically upwards in this case. All subjects performed this power test in three different positions corresponding to three phases in the lifting movement. Figure 82 shows these three positions and Figure 83 shows the results obtained from trials with sym-

167

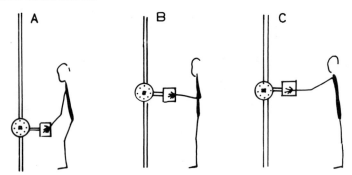

FIGURE 82. Study of lifting capacity on a dynamometer in three different positions.
(*See* further in the text.)

metrical lifts. A corresponding study was made of non-symmetrical left arm lifts.

Electromyographic studies of load on the back, neck, shoulder, and arm muscles showed that 'low' non-symmetrical lifts, i.e. those performed at about waist level (0·5 m), required the least muscular effort, even though the load rested on the left arm only. In 'high' non-symmetrical lifts, i.e. about chest level strain on the muscles was about as great as in symmetrical lifts at waist level. The heavy load on muscles in symmetrical lifts, lifts in which one might perhaps have expected the smallest muscular loading since load is evenly distributed on the right and left arms, is mainly due to abduction of the upper arm, always a part of these lifts. Another important cause of muscular exertion in the symmetrical lift is the forward extension of the arms which is part of these lifts. The load then acquires a very long lever with respect to the shoulder joint. As may be seen in Figure 83 the capacity of the muscles is only a little more than half of that available when the load is held in close to the body. This is also illustrated in Figure 84. Arm force was measured in three phases of the symmetrical lift. Activity in a few shoulder muscles was registered at the same time as arm power. Despite increased muscular exertion the external force produced was less.

'High' symmetrical lifts of loads of 4–5 kg, in particular if the lifts must be performed with outstretched arms, require so much muscular effort that they should be beyond the physio-

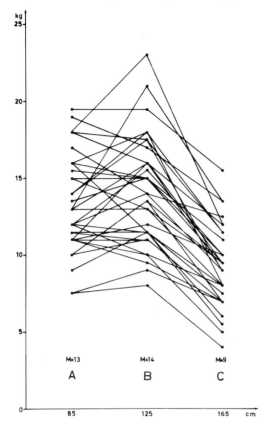

FIGURE 83. Lifting capacity in kilogrammes of 35 women in three different positions, A, B, and C. Distance between floor and dynamometer handle in centimetres.
(*See* further in the text.)

logical performance ability of many women. 'High' non-symmetrical lifts are also inappropriate. If high symmetrical lifts must be made in repetitive work, the lifting technique should be varied so that lifts are made with the right and left arms alternately.

## SLIPPING AND THE RISK OF SLIPPING IN THE FACTORY

Slipping is one of the most common causes of accidents on

169

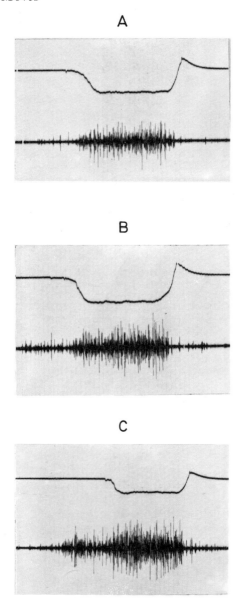

FIGURE 84. Recording of the lifting capacity and the muscle activity in right deltoid muscle in three different positions (corresponding to the positions A, B, and C in Figure 82).

170

the job. It is particularly in the food industry and food trade that such accidents are serious, elusive problems.

However, considerable efforts have been made in recent years to prevent these accidents. In this context, studies have been made of the friction between footwear and floor under various conditions. A consistent finding has been that the friction which exists prior to the slip is of critical importance for the occurrence of slipping. Thus, static friction is the critical value.

When two bodies come in contact, forces can be transmitted between them. These forces can be divided into two components, a normal force and a tangential force. These components act on both bodies to the same degree but in opposite directions, according to the law of action and reaction. Experience has shown that if the tangential component exceeds a certain value, the two bodies will start to slip. There is a definite relationship, known in physics as the coefficient of friction, between the maximum tangential force and a particular normal force. If the bodies start to slip, however, the tangential force does not generally display a constant relationship to the normal force because the coefficient of friction is then also related to the rate of slip. The frictional force may also be independent of the normal force if there is a liquid between the slip surfaces and if the rate of slip is low—only low rates are involved in the present context. In physics this is known as fluid friction and the frictional force is considered to be proportional to the rate of slip.

The tangential or frictional force between two sliding bodies is generally less than the force between the bodies immediately before they begin to slip. The frictional forces are called kinetic friction and static friction; static friction is greater than kinetic friction. Slipping occurs between the foot and the ground in walking when the tangential force exceeds the value for static friction. When slipping begins, the friction diminishes to kinetic friction and the foot continues to slip.

An estimate of the magnitude of tangential and normal forces in the habitual gait may be obtained from the power curves registered with the aid of force-plates. The appearance of such curves and the causes of their characteristic shape were described in greater detail in Chapter 5.

The gait of about 100 persons was studied for the special purpose of measuring the relationship between tangential force and normal force. It should first be pointed out that during these experiments none of the subjects slipped whether they walked barefooted, in their socks, or with their shoes on. Consecutive values of the vertical component ($V$), i.e. the normal force, can be obtained by measuring the distance between the zero-level of the $Z$ curve, Figure 85, and a series

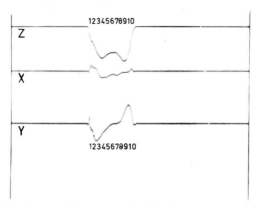

FIGURE 85. Calculation of the magnitude of the force components, particularly Z and Y, and their mutual relationships in ten different moments of the support phase in walking.

(*See* further in the text.)

of points on this during the support phase. Similar measurements of the corresponding horizontal tangential component can be made on the $Y$ and $Z$ curves. The relationship between the horizontal tangential and the vertical component, i.e. the relationship $H/V$, can then be calculated for the course of the support phase. The manner in which this relationship, i.e. the danger of slipping, varies during the support phase is illustrated in Figure 86.

There are two particularly critical instants, the first being when the foot touches the ground. The weight of the body is then lying behind the surface of contact between the foot and the ground, while the movement of the body's centre of gravity has just started the more rapid phase. The second critical instant is at the actual take-off, when the body's centre of gravity lies in front of the pushing foot. The first of these two

172

CASE C.S.

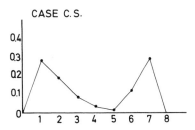

CASE R.B.

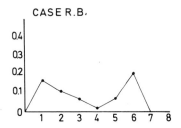

FIGURE 86. The relationship between the horizontal force and the vertical force, $H/V$, during the support phase in walking. Two subjects chosen at random.

instants, involves the greatest potential danger. If a person slips at this point, he is liable to fall backwards and he has then but a small chance to cushion his fall with his hands. If he slips at the second critical point he falls forward but in that case it is easier for him to break the fall with his hands.

There are naturally certain individual variations in the magnitude of the relationship $H/V$ and this also varies with the type of footwear. However, these variations are so small that no statistically significant difference could be demonstrated. The data in Figure 86 may therefore be regarded as representative of the habitual gait of a large majority of individuals.

The coefficient of friction between heel and floor must be greater than the highest value of the relationship $H/V$ during the support phase if slipping is to be avoided. The difference between this highest value and the coefficient of friction in question constitutes a measure of safety against slipping. The larger this difference, the smaller the risk of slipping. There must always be a certain difference as people in their daily work do not stick to the gait pattern registered in our laboratory experiment, i.e. walking in a straight line on level ground. The question of how large this difference should be cannot be determined until the relationship $H/V$ has been studied in sufficient detail at different places of work, regardless of whether accidents due to slipping occur there or not.

The value for $H/V$ obtained from walking experiments may be termed the requisite coefficient of friction $f_e$, for ordinary walking and may be regarded as the primary minimum requirement. If the coefficient of friction, $f$, between floor and foot-

173

wear is less than $f_e$, there is considerable risk of slipping. If $f_e$ is equal to $f$, the risk is still great unless the gait is changed, e.g. to shorter steps, but the risk declines rapidly as $f$ becomes greater than $f_e$. The mean value of $f_e$ in the present experiments was 0·23.

Methods for measuring friction between different types of floor and different materials for heels and soles have been described in the literature but all are unfortunately subject to a variety of shortcomings. In the first place, it is not at all clear which magnitudes characterize a risk of slipping and what the measurements imply in purely physical terms. In the second place, most of the measurements can only be made in the laboratory and not at places of work, while in the third place many of the methods do not give reproducible measurements and consequently call for statistical processing of a series of data.

Starting from the present calculation of the requisite co-efficient of friction, $f_e$, an instrument was constructed for measuring the coefficient of friction between a heel of, e.g. rubber, wood, or leather and the floor at different places of work, Figure 87. This instrument largely consists of a trolley

FIGURE 87. Friction meter.

with a 'leg' suspended in a low-friction joint on the underside of the trolley. The 'leg' is made of an aluminium rod the length of which is variable so that the 'leg' may form different angles to the perpendicular. The size of the angle is read off a graduated plate placed so that the 'leg's' joint axis passes through the centre of the plate. A heel of the material to be tested is fitted to the lower end of the 'leg'. The angle between the floor and the heel can also be varied. In the present experiments and tests, the angle was generally between 20 and 30°. This is the angle usually formed between the underside of the shoe and the floor in ordinary walking when the foot touches the ground, the most critical point in terms of slipping. Since the joint between the 'leg' and the trolley is practically free from friction, the trolley has no torque effect on the 'leg'. If one disregards the weight of the 'leg' itself, the force between the 'leg' and trolley in equilibrium always acts along the line of the 'leg'. This means that the force between the 'leg' and the floor in equilibrium also acts in the direction of the leg. Figure 88 clarifies this relationship.

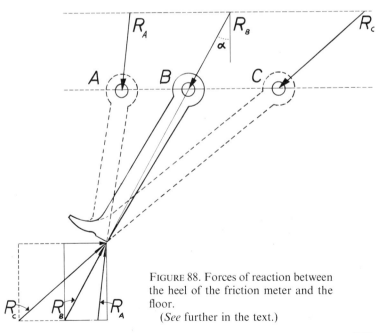

FIGURE 88. Forces of reaction between the heel of the friction meter and the floor.
(*See* further in the text.)

175

If a force is now brought to bear on the trolley, it will act via the joint of the 'leg' with a force, $R$. This force $R$ must be counteracted by an equally large force $R$ in the opposite direction from the floor. In case A the frictional force (the horizontal component of $R$) is smaller than in case B and in case B it is smaller than in case C.

If the horizontal component of $R_B$ is taken to be the maximum permissible frictional force for the vertical force in question, slipping will occur in case C but not in case A, where friction is not fully developed. The limiting angle, $\alpha$, at which the 'leg' just does not slip on the floor, i.e. when slipping would occur at a slightly larger angle, can be used to derive the coefficient of friction required, $f = \tan \alpha$.

The apparatus was used to study floor friction in a meat factory. Only friction between rubber heels and the floor was tested in this study. As will be seen from Table III the coefficient of friction varied between 0·10 and 0·64. The risk of

TABLE III

Coefficients of friction between rubber heel and floor in a meat factory, as obtained in tests with a friction meter

| *Place* | $\alpha°$ | $f$ | *Evaluation* |
|---|---|---|---|
| Outside cooking room | 7·5 | 0·13 | Major risk of slipping |
| In front of preserving table | 15·5 | 0·27 | Risk of slipping |
| In front of preserving table | 14·0 | 0·24 | Risk of slipping |
| In front of preserving table | 6·0 | 0·10 | Major risk of slipping |
| In front of preserving table | 16·0 | 0·28 | Risk of slipping |
| In front of preserving table | 22·0 | 0·40 | No risk of slipping |
| Near mixer | 24·0 | 0·44 | No risk of slipping |
| Near mixer | 33·0 | 0·64 | No risk of slipping |
| Entrance to main kitchen | 28·0 | 0·53 | No risk of slipping |

slipping was very great for parts of the floor area with a coefficient of friction less than 0·30 and practically non-existent when 0·50 or more. These variations reflect the type of floor material as well as the degree and nature of soiling of the floor. The floor was dry in some places and covered with water, animal refuse, or fat in others. It was difficult to determine whether it was the type of floor or the layer of fat or water covering the floor which primarily determined the size of the coefficient of friction.

In order to obtain more controlled experimental conditions, a series of measurements were made in a department for food research which had a glazed tile floor. First of all, tests were made on the dry floor. A second series was made after the floor had been hosed down with water. Friction tests were also made after brawn jelly, glycerine and horse fat had been spread out over the floor. The results are shown in Table IV from which it will be seen that a floor which is relatively safe when dry and clean may be extremely dangerous when soiled.

## MANUAL AND ELECTRICAL TYPEWRITERS

Five experienced typists took part in a comparison test between typing on a manual and on an electrical typewriter. Each typed the same text eight times, four times on each machine. The text was a 440 word long newspaper article. A strain gauge was built into the 'N' key of each typewriter to register the force of the right forefinger against the key at each stroke. 'N' was chosen by drawing lots among the most common letters of the alphabet. Subjects were naturally unaware that there was such a sensor.

The time integral or impulse exerted by the finger was used as a comparative value. Figure 89 shows the appearance of the power curves registered each time the 'N' key was struck. The area defined by the power curve and base line, indicates the magnitude expressed in Newton-seconds.

FIGURE 89. Registration of finger force in typing with an electric typewriter: paper speed = 100 mm/s.
(*See* further in the text.)

The power curves registered in typing with the electric machine displayed a little twitch in their ascending phase. This twitch developed as soon as the key passed the critical position in which the key's lever system electrically triggered

177

TABLE IV
Coefficients of friction between rubber and wooden heels and a floor soiled by different substances

| Floor | WOODEN HEEL | | | RUBBER HEEL | | |
|---|---|---|---|---|---|---|
| | $\alpha°$ | $f$ | Evaluation | $\alpha°$ | $f$ | Evaluation |
| Dry | 16·6 | 0·30 | Risk of slipping | 33·8 | 0·65 | No risk of slipping |
| With layer of water | 24·6 | 0·46 | No risk of slipping | 26·6 | 0·50 | No risk of slipping |
| Smeared with brawn jelly | 17·5 | 0·32 | Risk of slipping | 14·0 | 0·25 | Risk of slipping |
| Smeared with thicker layer of brawn jelly | 14·0 | 0·25 | Risk of slipping | 9·0 | 0·16 | Major risk of slipping |
| Moistened with glycerine | 14·5 | 0·26 | Risk of slipping | 7·0 | 0·12 | Major risk of slipping |
| Smeared with minced horse fat | 13·0 | 0·23 | Risk of slipping | 7·0* | 0·12 | Major risk of slipping |

*The spread was so wide that the mean value cannot be regarded as representative. Therefore, only the smallest value measured is given here.

the type bar. Thus, the vertical distance between the base line and the twitch indicates the force required for a stroke and the area to the left of this vertical distance indicates the effort required to depress a key enough to close the electrical circuit. But all work to the right of the vertical distance was unnecessary.

Thus, the greater part of the finger's total work effort is superfluous work. If the key is depressed very slowly and cautiously the finger's movement can be stopped at the critical position. But this is impossible with rapid finger movements because the kinetic energy of the fingers is so great that their movement does not terminate until stopped by external mechanical resistance, i.e. when the key bottoms. This is the main reason why the ratio between the fingers' work performance in typing with manual and with electric typewriters was not greater than about 2·5 in these studies, as may be seen in Table V.

TABLE V

| Subject | Mean impulse value*, mm² | | Ratio manual/electrical typewriter |
|---|---|---|---|
| | Electrical typewriter | Manual typewriter | |
| A | 40 ± 5 | 103 ± 6 | 2·58 |
| B | 67 ± 3·5 | 98 ± 4 | 1·47 |
| C | 49 ± 4 | 129 ± 8·5 | 2·63 |
| D | 43 ± 5 | 112 ± 10 | 2·60 |
| E | 49·5 ± 3·5 | 120 ± 6 | 2·43 |

*Impulse = time integral of the force.

In one subject, B, there was scarcely any difference at all. This was mainly because this typist has used a manual typewriter only for 26 years and had only gone over to an electrical machine two years prior to the test. In the course of 26 years she had developed an innervation pattern for her finger movements, a pattern which had become so fixed that it could not be influenced significantly by the relatively big differences in resistance presented by the keys of the two typewriters to her fingers. In other words she was less able to optimize her finger forces than other subjects, as was also clear from the electromyograms. The activity of her finger flexors was quantitatively the same, irrespective of the typewriter used. Activity was considerably less in the other four typists when they used the electric typewriter than when typing with the manual typewriter.

It is naturally difficult to draw any general conclusions from this study. But it may serve to illustrate the manner in which our movements are controlled by both neuromuscular and mechanical–anatomical laws. If one of the afferent impulses upon which a movement pattern is based, e.g. the mechanical resistance of the keys, is changed then the efferent flow of impulses to muscles is also changed. This is also true even in movement patterns as subtle and well-established as finger movements in typing.

However, adaptation is not complete. The study showed that the nervous system is unable to adapt finger power to the resistance fingers must overcome in movements as rapid as those in question. This inability to optimize causes the finger muscles to perform approximately 100 times more work than is necessary in typing with an electric typewriter. A person who had never used a manual typewriter but has learnt to type with an electric typewriter from the beginning might conceivably have a better chance of reducing superfluous effort than a person who had learnt to type using a manual typewriter, i.e. a machine requiring great finger power, for several years.

# Bibliography

ÅKERBLOM, B. *Standing and sitting posture.* Nordiska Bokhandeln, Stockholm, 1948.

ASMUSSEN, E. The weight-carrying function of the human spine. *Acta orthop. scand.*, **29**: 276–90, 1960.

ASMUSSEN, E. and KLAUSEN, K. Form and function of the erect human spine. *Clin. orthop.*, **25**: 55–63, 1962.

ASMUSSEN, E. The biological bases of sport. *Ergonomics*, vol. **8**, no. 2, 1965.

BÄCKDAHL, M. and CARLSÖÖ, S. Distribution of activity in muscles acting on the wrist. *Acta morphol. neerl.-scand.*, vol. **IV**, no. 2, 1960.

BACKHOUSE, K. M. and CATTON, W. T. An experimental study of the functions of the lumbrical muscles in the human hand. *J. anat.*, **88**: 133–41, 1954.

BARKLA, D. The estimation of body measurements of British population in relation to seat design. *Ergonomics*, vol. **4**, no. 2, 1961.

BARNES, R. M. *Time and motion study.* John Wiley and Sons Inc., U.S.A., 1949.

BASMAJIAN, J. V. and BENTZON, J. W. An electromyographic study of certain muscles of the leg and foot in the standing position. *Surg. gynec. & obst.*, **98**: 662–6, 1954.

BASMAJIAN, J. V. Electromyography of iliopsoas. *Anat. rec.*, **132**: 127–32, 1958.

BASMAJIAN, J. V. and TRAVILL, A. Electromyography of the pronator muscles in the forearm. *Anat. rec.*, **139**: 45–9, 1961.

BASMAJIAN, J. V., FORREST, W. J., and SHINE, G. A simple connector for fine-wire electrodes. *J. appl. physiol.*, **21**: 1680, 1966.

BASMAJIAN, J. V. *Muscles alive, their functions revealed by electromyography*. The Williams & Wilkins Co., Baltimore, second edition, 1967.

BAUWENS, P. Electromyography. *Brit. j. phys. med.*, **11**: 130–6, 1948.

BEARN, J. G. The significance of the activity of the abdominal muscles in weight lifting. *Acta anat.*, **45**: 83–9, 1961.

BEARN, J. G. An electromyographic study of the trapezius, deltoid, pectoralis major, biceps and triceps muscles, during static loading of the upper limb. *Anat. rec.*, **140**: 103–8, 1961.

BIERMAN, W. and RALSTON, H. J. Electromyographic study during passive and active flexion and extension of the knee of the normal human subject. *Arch. phys. med.*, **46**: 71 5, 1965.

BIGLAND, B. and LIPPOLD, O. C. J. The relation between force, velocity and integrated electrical activity in human muscles. *J. physiol.*, **123**: 214–24, 1954.

BIGLAND, B. and LIPPOLD, O. C. J. Motor unit activity in the voluntary contraction of human muscle. *J. physiol.*, **125**: 322–35, 1954.

BINKHORST, R. A. and CARLSÖÖ, S. The thumb–forefinger grip and the shape of handles of certain instruments: An electromyographic study of the muscle-load. *Ergonomics*, vol. **5**, no. 3, 467, 1962.

BJÖRK, G. Studies on the draught force of horses. *Acta agriculturae scand.*, suppl. **4**, 1958.

BOUISSET, S., DENIMAL, J., and SOULA, C. Relation entre l'accélération d'un raccourcissement musculaire et l'activité électromyographique intégrée. *J. physiol.*, Paris, **55**: 203, 1963.

BRACE, D. K. *Measuring motor ability*. Barnes, New York, 1927.

BRAUNE, W. and FISCHER, O. *Der Gang des Menschen*. Abh. d. Mathem.–Phys. Klasse d. Kgl. Sächs. Ges. d. Wiss., Leipzig. S. Hirzl, 1895.

BUCHTHAL, F. The functional organization of the motor unit: a summary of results. *Am. j. phys. med.*, **38**: 125–8, 1959.

BUNN, J. W. *Scientific principles of coaching*. Englewood Cliffs, N.J., Prentice-Hall Inc., 1955

CARLSÖÖ, S. Eine elektromyographische Untersuchung der Muskelaktivität im Musculus Deltoideus. *Acta morphol. neerl.-scand.*, vol. **II**, no. 4, 1959.

CARLSÖÖ, S. The mechanics of the two-joint muscles rectus femoris, sartorius and tensor fasciae latae in relation to their activity. *Scand. j. rehab. med.*, **1**: 107–11, 1969.

CARLSÖÖ, S. The static muscle load in different work positions: an electromyographic study. *Ergonomics*, **4**: 193–211, 1961.

CARLSÖÖ, S. A method for studying walking on different surfaces. *Ergonomics*, vol. **5**, no. 1, 271, 1962.

CARLSÖÖ, S. and JOHANSSON, O. Stabilization of a load on the elbow joint in some protective movements. *Acta anat.*, **48**: 224–31, 1962.

CARLSÖÖ, S. Schreiben mit mechanischer und elektrischer Schreibmaschine. *Proceedings of the 2nd I.E.A. Congress, 1964*, p. 361.

CARLSÖÖ, S. A kinetic analysis of the golf swing. *The journal of sports medicine and physical fitness.* vol. **7**, no. 2, 76–82, 1967.

CARLSÖÖ, S. and NORDSTRAND, A. The coordination of the knee-muscles in some voluntary movements and in the gait in cases with and without knee joint injuries. *Acta chir. scand.*, **134**: 423–6, 1968.

CARLSÖÖ, S. and WETZENSTEIN, H. Change of form of the foot and the foot skeleton upon momentary weight-bearing. *Acta orthop. scand.*, **39**: 413–23, 1968.

CHAPMAN, M. W. and RALSTON, H. J. Effect of immobilization of the back and arms on energy expenditure during level walking. Vocational rehabilitation administration research grant RD–1112–M, no. 52, 1964.

DEMPSTER, W. T. and FINERTY, J. C. Relative activity of wrist-moving muscles in static support of the wrist joint: an electromyographic study. *Am. j. physiol.*, **150**: 596–606, 1947.

DUCHENNE, G. B. *Physiology of motion.* Original title: *Physiologie des mouvements.* Translated into English by E. B. Kaplan. W. B. Saunders Co., Philadelphia and London, 1959.

EBBERHART, H. D., INMAN, V. T., SAUNDERS, J. B. deC. M., LEVENS, A. S., BRESLER, B., and COWAN, T. D. *Fundamental* studies of human locomotion and other information related to the design of artificial limbs. A report to the N.R.C. Committee on artificial limbs. University of California, Berkeley, 1947.

EBBERHART, H. D., INMAN, V. T., and BRESLER, B. The principal elements in human locomotion. In *Human limbs and their substitutes*. Ed. by P. E. Klopsteg and P. D. Wilson. McGraw-Hill Book Co., New York, 1954.

ELFTMAN, H. The function of the arms in walking. *Human Biol.*, **11**: 529–35, 1939.

ELFTMAN, H. The basic pattern of human locomotion. *Ann. New York Acad. Sc.*, **51**: 1207, 1951.

ELFTMAN, H. Biomechanics of muscle: with particular application to studies of gait. *J. Bone & Joint Surg.*, **48-A**: 363–77, 1966.

FLOYD, W. F. and SILVER, P. H. S. Electromyographic study of patterns of activity of the anterior abdominal wall muscles in man. *J. anat.*, **84**: 132–45, 1950.

FLOYD, W. F. and SILVER, P. H. S. Function of erector spinae in flexion of the trunk. *Lancet*, Jan. 20, 133–8, 1951.

FLOYD, W. F. and SILVER, P. H. S. The function of the erector spinae muscles in certain movements and postures in man. *J. physiol.*, **129**: 184–203, 1955.

FROST, H. M. *An introduction to biomechanics*. Charles C. Thomas, Springfield, Illinois, U.S.A., 1967.

GESELL, A. *The first five years of life*. Methuen and Co., London, 1963.

GRIFFITHS, H. E. Treatment of the injured workman. *Lancet*, **I**, 729, 1943.

HALL, M. C. *The locomotor system: Functional anatomy*. Charles C. Thomas, Springfield, Illinois, U.S.A., 1965.

HARDY, R. H. A method of studying muscular activity during walking. *Med. & Biol. Illustration*, **9**: 158–63, 1959.

HELLEBRANDT, F. A. The physiology of motor learning. *Cereb. Palsy Rev.*, **10**: 9, 1958

HICKS, J. H. The plantar aponeurosis and the arch. *J. anat.*, London, **88**: 25, 1954.

HICKS, J. H. The foot as a support. *Acta anat.*, **25**: 34, 1955.

HICKS, J. H. The action of muscles on the foot in standing. *Acta anat.*, **27**: 180, 1955.

HOUTZ, S. J. and FISCHER, F. J. An analysis of muscle action and joint excursion during exercise on a stationary bicycle. *J. Bone & Joint Surg.*, **41-A**: 123–31, 1959.

HOUTZ, S. J. and WALSH, F. P. Electromyographic analysis of the function of the muscles acting on the ankle during weight-bearing with special reference to the triceps surae. *J. Bone & Joint Surg.*, **41-A**: 1469–81, 1959.

HOUTZ, S. J. and FISCHER, F. J. Function of leg muscles acting on the foot as modified by body movements. *J. appl. physiol.*, **16**: 597–605, 1961.

HOWARD, J. P. and TEMPLETON, W. B. *Human spatial orientation*. John Wiley and Sons, London, New York, Sydney, 1966.

INMAN, V. T., SAUNDERS, J. B. deC. M., and ABBOTT, L. C. Observations on the function of the shoulder joint. *J. Bone & Joint Surg.*, **26**: 1–30, 1944.

JOHNSON, W. R. *Science and medicine of exercise and sports*. Harper and Bros., New York, 1960.

JONES, F. WOOD. *The principles of anatomy as seen in the hand*. Second edition. Baillière, Tindall & Cox, London, 1941.

JONSSON, B. and STEEN, B. Function of the gracilis muscle. An electromyographic study. *Acta morph. neerl.-scand.*, **5**: 269, 1963.

JONSSON, B. and KARLSSON, E. Function of the gluteus maximus muscle: an electromyographic study. *Acta morphol. neerl.-scand.*, **6**: 161–9, 1965.

JONSSON, B. The functions of individual muscles in the lumbar part of the spine muscle. *Electromyography*, vol. **10**, no. 1, 5–21, 1970.

JOSEPH, J. and NIGHTINGALE, A. Electromyography of muscles of posture: leg muscles in males. *J. physiol.*, **117**: 484–91, 1952.

JOSEPH, J. and NIGHTINGALE, A. Electromyography of muscles of posture: thigh muscles in males. *J. physiol.*, **126**: 81–5, 1954.

JOSEPH, J. and NIGHTINGALE, A. Electromyography of muscles of posture: leg and thigh muscles, including the effects of high heels. *J. physiol.*, **132**: 465–8, 1956.

JOSEPH, J. and WILLIAMS, P. L. Electromyography of certain hip muscles. *J. anat.*, **91**: 286–94, 1957.

185

JOSEPH, J. *Man's posture: Electromyographic studies.* Charles C. Thomas, Springfield, Illinois, U.S.A., 1960.

KARPOVICH, P. V. and WILKLOW, L. B. A goniometric study of the human foot in standing and walking. *United States Armed Forces medical journal*, vol. **X**, no. 8, 1959.

KATZ, B. *Nerve, muscle and synapsa.* McGraw-Hill Book Co., London, 1966.

KEEGAN, J. J. Alterations of the lumbar curve. *J. Bone & Joint Surg.*, vol. **35-A**: no. 3, 1953.

KELLERMAN, F. TH., WALY, P. A., and WILLEMS, P. J. *Mensch und Arbeit in der Industrie*, Philips Technische Bibliotek, 1964.

KNUTSSON, B., LINDH, K., and TELHAG, H. Sitting—an electromyographic and mechanical study. *Acta orthop. scand.*, vol. **37**, fasc. 4, 1966.

LANDSMEER, J. M. F. The coordination of finger-joint motions. *J. Bone & Joint Surg.*, **54-A**: 1963.

LIPPOLD, O. C. J. The relation between integrated action potentials in a human muscle and its isometric tension. *J. physiol.*, **117**: 492–9, 1952.

LONG, C., BROWN, M. E., and WEISS, G. An electromyographic study of the extrinsic–intrinsic kinesiology of the hand: preliminary report. *Arch. phys. med.*, **41**: 175–81, 1960.

LONG, C., BROWN, M. E., and WEISS, G. Electromyographic kinesiology of the hand. Part II. Third dorsal interosseus and extensor digitorum of the long finger. *Arch. phys. med.*, **42**: 559–65, 1961.

LONG, C. and BROWN, M.E. Electromyographic kinesiology of the hand. Part III. Lumbricalis and flexor digitorum profundus to the long finger. *Arch. phys. med.*, **43**: 450–60, 1962.

LONG, C. and BROWN, M. E. Electromyographic kinesiology of the hand: muscles moving the long finger. *J. Bone & Joint Surg.*, **46-A**: 1683–706, 1964.

MCBRIDE, E. D. *Disability evaluation.* Third edition. J. B. Lippincott Co., Philadelphia, London, Montreal, 1942.

MOLBECK, S. On the paradoxial effect of some two-joint muscles. *Acta morphol. neerl.-scand.*, **6**: 171, 1965.

MORRIS, J. M., LUCAS, D. B., and BRESLER, B. *The role of the trunk in stability of the spine.* Biomechanics laboratory, University of California, Publications no. 42, 1961.

MORRIS, J. M., BRENNER, G., and LUCAS, D. B. An electromyographic study of the muscles of the back in man. *J. anat.*, **96**: 509–20, 1962.

MORTON, D. J. *Human locomotion and body form. A study of gravity and man.* The Williams & Wilkins Co., Baltimore, 1952.

NACHEMSON, A. Electromyographic studies of the vertebral portion of the psoas muscle. *Acta orthop. scand.*, **37**: 177–90, 1966.

NACHEMSON, A. and ELFSTRÖM, G. Intravital dynamic pressure measurements in lumbar discs. A study of common movement maneuvers and exercises. *Scandinavian Journal of Medicine*, suppl. 1, 1970.

NAPIER, J. R. The prehensile movements of the human hand. *J. Bone & Joint Surg.*, vol. **38-B**: no. 4, 1956.

O'CONNELL, A. L. and GARDNER, E. B. The use of electromyography in kinesiological research. *Res. Quart.*, **34**: 166–84, 1963.

PAULY, J. E. An electromyographic analysis of certain movements and exercises. Part I: some deep muscles of the back. *Anat. rec.*, **155**: 223–34, 1966.

PORTNOY, H. and MORIN, F. Electromyographic study of postural muscles in various positions and movements. *Am. j. physiol.*, **186**: 122–6, 1956.

RADCLIFFE, C. W. and RALSTON, H. J. *Performance characteristics of fluid-controlled prosthetic knee mechanisms.* Berkeley, San Francisco, no. 49, 1963.

RALSTON, H. J. Uses and limitations of electromyography in the quantitative study of skeletal muscle function. *Am. j. orthodont.*, **47**: 530–41, 1961.

RALSTON, H. J. Effects of immobilization of various body segments on the energy cost of human locomotion. *Ergonomics*, **7**: 53–60, 1964.

SAHA, A. K. Surgery of the paralysed and flail shoulder. *Acta orthop. scand.*, suppl. 97, 1967.

SAUNDERS, J. B. deC. M., INMAN, V. T., and EBBERHART, H. D. The major determinants in normal and pathological gait. *J. Bone & Joint Surg.*, **35-A**: 538–43, 1953.

SCHOBERTH, S. *Sitzhaltung, Sitzschaden, Sitzmöbel.* Springer-Verlag, Berlin—Göttingen—Heidelberg, 1962.

SCHWARTZ, R. P., HEATH, B. S., MORGAN, W., and TOWNA, C. A quantitative analysis of recorded variables in the walking pattern of 'normal' adults. *J. Bone & Joint Surg.*, vol. **1**, **46-A**: no. 2, 324–34, 1964.

SEYFFARTH, H. *The behaviour of motor-units in voluntary contraction.* A. W. Bröggers Boktrykkeri A/S, Oslo, 1940.

SKOGLUND, S. The activity of muscle receptors in the kitten. *Acta phys. scand.*, vol. **50**, 1960.

SMITH, K. U. and SMITH, W. M. *Perception and motion.* W. P. Saunders Co., Philadelphia and London, 1962.

STEINDLER, A. *Kinesiology of the human body under normal and pathological conditions*, Charles C. Thomas, Springfield, Illinois, U.S.A., 1955.

SUTHERLAND, D. H. An electromyographic study of the plantar flexors of the ankle in normal walking on the level. *J. Bone & Joint Surg.*, **48-A**: 66–71.

THOMAS, D. P. and WHITNEY, R. J. Postural movements during normal standing in man. *J. anat.*, London, **93**: 524–39, 1959.

WARTENWEILER, J., JOKL, E., and HEBBELINCK, M. *Biomechanics.* Proceedings of the First international seminar on biomechanics, Zürich, August 21–3, 1967. S. Karger, Basel, 1968.

WEIL, S. and WEIL, U. H. *Mechanik des Gehens.* Georg Thieme Verlag, Stuttgart, 1966.

WETZENSTEIN, H. Eine Untersuchung der Fersenbelastung beim Gehen. Eine Methode für die Messung der Fersenbelastung im Schuh. *Acta orthop. scand.*, suppl. no. 75, 1964.

WHITNEY, R. J. The stability provided by the feet during manoeuvres whilst standing. *J. anat.*, vol. **96**, part 1, 1962.

WHITSETT, C. E. *Some dynamic response characteristics of weightless man.* Technical documentary report no. AMRL–TDR–63–18, 1963. Behavioral Science Laboratory, 6570th Aerospace Medical Research Lab.

WILLIAMS, M. and LISSNER, H. R. *Biomechanics of human motion.* W. B. Saunders Co., Philadelphia, London, 1962.

# Glossary

*abduction*, move outwards away from middle line.

*acceleration, angular*, the change in angular velocity per unit of time.

*action potential*, the electrical effect observed in a muscle or nerve during activity.

*adduction*, move inwards towards central axis of the body.

*afferent*, conducting inwards.

*agonist*, a muscle which actively participates in and facilitates movement.

α (*alpha*) *motor neurone*, motor nerve cell.

*ante-torsion*, forward twisting.

*anthropometry*, study of the comparative measurements of the human body's proportions.

*aponeurosis*, a broad membranous tendon at the end of a muscle.

*axial*, on or along the axis.

*axon*, the impulse-carrying process of a typical nerve cell.

*basal ganglia*, three large accumulations of nerve cells in the cerebrum.

*bilateral*, relating to or having two sides.

*biomechanics*, the science of the application of mechanics to living creatures.

*caudal*, directed downwards.

*cerebellum*, concerned with the co-ordination of muscle action in the performance of movement or maintenance of posture.

*chronocyclography*, series of exposures at short and known intervals on the same negative.

*co-contraction*, simultaneous activation of an agonist and an antagonist.

*component*, each individual force included in a system of forces.

*co-ordination*, the harmonious mutual adjustment of parts so that they combine to carry out a function.

*cranio-caudal*, in a direction from the head towards the feet (actually from head to tail).

*depolarization*, opposite of polarization (q.v.).

*deviation*, turning away from the normal position or course.

*diphasic curve*, occurring in two stages.

*distal*, refers to a point on the extremities which is farther from the trunk than another. Its opposite is proximal.

*dorsal*, towards the back (opposite of ventral).

*dynamometer*, an instrument which measures the power of contraction of muscles.

*efferent*, conducting outwards.

*electrogoniography*, technique to measure the movement of a joint in terms of its direction, velocity, and acceleration.

*electromyography*, registration of electrical activity in muscles.

*equivalent*, equal in any respect.

*ergonomics*, the science of man and his work.

*eversion*, pronation. In eversion the outer foot margin is raised at the same time as the toe-tips are moved outwards and upwards somewhat.

*extension*, the act of extending a joint.

*extensors*, muscles which extend a joint.

*exteroceptive impulses*, impulses caused by stimuli from without.

*flexion*, the act of bending a joint.

*flexors*, muscles which flex a joint.

*force-plate meter*, used to measure forces exerted by the feet against the ground.

*γ (gamma) neurone*, motor nerve cell whose axon terminates in a muscle spindle.

*ganglia*, a collection or accumulation of nerve cells.

*ginglymus joint*, a hinge joint.

*gradient*, fall in electrical potential.

*impedance*, electrical resistance in an alternating current circuit.

*impulse*, the time element of power.

*innervation pattern*, complex system of nerve cells linked together in chains passing through various parts and levels of the brain and spinal cord.

*insertion site*, point of attachment.

*intercostal*, between the ribs.

*intermittent*, repeated at certain intervals.

*inter-neurone*, a nerve cell connecting two other nerve cells.

*interstitial*, relating to an interspace, e.g. between cells and cell groups.

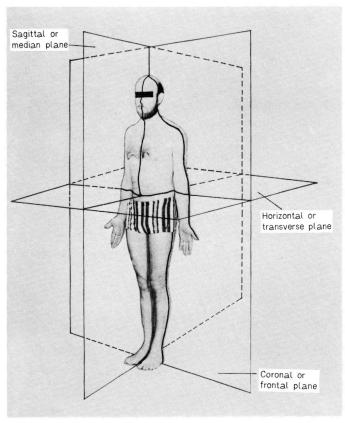

Sagittal or median plane

Horizontal or transverse plane

Coronal or frontal plane

Diagram showing the planes of the body.

*intervertebral*, between spinal vertebrae.

*intra-abdominal*, in the abdominal cavity.

*intra-thoracic*, in the chest cavity.

*isometric*, muscular contraction with muscle length unchanged.

*isotonic*, muscular contraction with muscle length shortened.

*kinetics* and *kinetic analysis*, branch of science dealing with the motion produced in bodies by forces acting upon them.

*kinetic energy*, the energy possessed by a moving body by virtue of its motion. The formula for kinetic energy is $\frac{1}{2}mv^2$ erg or $mv^2/2g$ kilogramme-metre.

*kyphosis*, curvature of the spine with a backward convexity.

*lateral*, located at the side.

*lipoid*, fatty.

*lordosis*, curvature of the spine with a forward convexity.

*lumbar*, relating to the small of the back.

*luxation*, dislocation.

*medial*, relating to the middle or centre.

*medial plane*, middle plane.

*moment of momentum* (*see* momentum, angular).

*momentum*, the product of the mass of a body and its linear velocity.

*momentum, angular*, the product of the moment of inertia and a body's angular velocity.

*motor end-plate*, connection between a motor nerve and a muscle fibre.

*motor-unit*, all of the muscle fibres supplied by a single motor nerve cell.

*muscle receptor*, sensory organ in muscles.

*myogram* (*see* electromyography—EMG).

*nerve end-plate*, motor end-plate.

*neurone*, a nerve cell.

*normal force*, force perpendicular to a surface (*see* tangential force).

*optimize*, to do in the most favourable way possible.

*permeability*, diffusion of a gas or liquid through a porous material.

*phylogenesis*, evolutionary development of the species, as opposed to ontogenesis (the development of the individual).

*plantar*, relating to the sole of the foot.

*polarization*, formation of two poles.

*postural*, relating to body posture.

*postural sway*, minor, subconscious body motions performed when a person is apparently standing still.

*projection system*, paths of conduction between the cerebral cortex and other parts of the central nervous system.

*pronate*, to turn inwards or forwards.

*pronation* (*see* eversion).

*proprioceptive impulses*, impulses which through stimulation of sensor organs located in muscles, ligaments, joints, etc., provide information on the body's position and movements.

*proximal* (*see* distal).

*radial*, relating to the radius bone.

*radian*, a unit of circular measure which is equal to approximately 57·2°.

*receptors*, organs for receiving sensory impressions.

*reponate*, to replace or reset.

*reposition*, to reset in a correct position.

*resistance*, electrical resistance in a direct current circuit.

*retro-torsion*, twisting backwards.

*sagittal plane*, a vertical plane through the body and perpendicular to the frontal plane.

*sensory*, relating to sensation.

*supinate*, to bend backwards or outwards; opposite of pronate.

*syndesmosis*, a joint with joint surfaces lined by fibrous connective tissue.

*synergist group*, muscles with the same mechanical conditions.

*tangential force*, force applied along a surface (*see* normal force).

*tendon receptor*, (also called a Golgi receptor) a sensory organ in tendons.

*topographical*, mutual positional relationship.

*translation*, a movement in which all the points on a body describe straight or curved parallel lines.

*transverse*, cross-wise.

*ulnar*, towards the ulnar or little finger side.

*ventral*, towards the abdomen (opposite of dorsal).
*vestibulum*, central cavity of the ear and part of the apparatus
for maintaining body balance.
*volar*, palmar.
*voluntary*, performed by an exercise of will.

# Index